Introductory Guide to
High-Performance Audio Systems

Stereo • Surround Sound • Home Theater

Robert Harley

Acapella Publishing
P.O. Box 1768
Tijeras, NM 87059
hifibooks.com

International Standard Book Number:
0-9786493-0-3
978-0-9786493-0-2

Cover Concept and Design: *Torquil Dewar*
Cover image: Bowers & Wilkins (B&W) 800D loudspeaker
Illustrations: *Torquil Dewar, Rick Velasquez* and *Nancy Josephson,* except where
otherwise credited

Printed in the United States of America

Contents

Contents

Contents

Contents

Contents

Foreword

This book was inspired by my passion to tell everyone who loves music that the listening experience can be more engaging, involving, and pleasurable through a high-performance audio system. It's a well-kept secret to the general public that there's a whole category of audio equipment that can reproduce music with a depth, intensity, and listener involvement that mass-market audio products don't begin to match. It remains a secret because the high-end audio industry, to its detriment, has focused on the hard-core audio enthusiasts rather than appealing to a broader audience of listeners who simply love music and want to hear it reproduced with the highest fidelity.

My goal for this book is to make high-performance audio easily understandable and accessible to anyone who enjoys music.

I approached this task with some fundamental beliefs:

1) Music is vitally important to the quality of our lives.

2) An audio system is a vehicle for exploring the world of music.

3) The better the quality of the reproduction, the deeper the listener's connection with the music.

4) Anyone can appreciate the difference between mediocre and superb sound reproduction.

5) A true high-end system can be assembled for not much more—and often less—than a mass-market system.

6) High-end audio should be accessible to anyone who loves music, not just technophiles or the wealthy.

I'd hoped that my first book, *The Complete Guide to High-End Audio*, would help everyone who read it to better connect with music. But at 640 pages, and with a fair amount of technical depth, it didn't quite fulfill the mission of making high-end audio accessible to ordinary music lovers.

That's where *Introductory Guide to High-Performance Audio Systems* comes in. In it, I've reworked much of the basic material from *The Complete Guide to High-End Audio* into a simpler, easier-to-digest form. I've also included sections on multichannel audio for music and home theater on the assumption that many of you will use a multichannel audio system for stereo music, multichannel music, and home theater. This new book will, I hope, reach many more music lovers, pointing them in directions that allow them to enjoy their favorite music wonderfully reproduced night after night.

Robert Harley
January, 2007

About the Author

Robert Harley is the Editor-in-Chief of *The Absolute Sound* and the former Editor-in-Chief of *The Perfect Vision* magazines. *The Absolute Sound*, founded in 1972, is the world's most respected journal of high-end audio. He is the author of two best-selling books in the field, *The Complete Guide to High-End Audio* and *Home Theater for Everyone: A Practical Guide to Today's Home-Entertainment Systems*. *The Complete Guide to High-End Audio* (now in a third edition) is widely considered the reference book in high-performance music reproduction, and has sold more than 100,000 copies in four languages.

Robert Harley holds a degree in audio engineering, and has taught a college degree program in that field. He has worked as a recording engineer and studio owner, compact disc mastering engineer, technical writer, and consumer-electronics journalist. He has written more than 700 published product reviews and articles on audio and home theater. Before joining *The Absolute Sound* and *The Perfect Vision* in 1999, he was Technical Editor of *Stereophile* magazine for eight years, and also served in that capacity at *Fi: The Magazine of Music and Sound*.

What is High-End Audio?

igh-end audio is about passion—passion for music, and for how well it is reproduced. High-end audio is the quest to re-create in the listener's home the musical message of the composer or performer with the maximum realism, emotion, and intensity. Because music is important, re-creating it with the highest possible fidelity is important.

High-end audio products constitute a unique subset of music-reproduction components that bear little similarity to the "stereo systems" sold in department stores. A music-reproduction system isn't a home appliance like a washing machine or toaster; it is a vehicle for expressing the vast emotional and intellectual potential of the music encoded on our records and CDs. The higher the quality of reproduction, the deeper our connection with the music.

The high-end ethos—that music and the quality of its reproduction matter deeply—is manifested in high-end audio products. They are designed by dedicated enthusiasts who combine technical skill and musical sensitivity in their crafting of components that take us one step closer to the original musical event. High-end products are designed by ear, built by hand, and exist for one reason: to enhance the experience of music listening.

A common misperception among the hi-fi–consuming public is that high-end audio means high-*priced* audio. In the mass-market mind, high-end audio is nothing more than elaborate stereo equipment with fancy features and price tags aimed at millionaires. Sure, the performance may be a little better than the hi-fi you find at your local appliance store, but who can afford it? Moreover, high-end audio is seen as being only for trained, discriminating listeners, snobs, or gadget freaks—but not for the average person on the street.

High-end audio is none of these things.

First, the term "high-end" refers to the products' *performance*, not their price. Many true high-end systems cost no more—and often less—than the all-in-one rack systems sold in department stores. I've heard many inexpensive systems that capture the essence of what high-quality music reproduction is all about—systems easily within the budgets of average consumers. Although many high-end components *are* high-priced, this doesn't mean that you have to take out a second mortgage to have high-quality music reproduction in your home. A great-sounding system can be less expensive than you might think.

Second, high-end audio is about communicating the musical experience, not adding elaborate, difficult-to-operate features. In fact, high-end systems are much easier to use than mass-market mid-fi systems. This is because the high-end ethic eliminates useless features, instead putting the money into sound quality. High-end audio is for music lovers, not electronics whizzes.

Chapter 1

Third, *anyone* who likes music can immediately appreciate the value of high-quality sound reproduction. It doesn't take a "golden ear" to know what sounds good. The differences between good and mediocre music reproduction are instantly obvious. The reaction—usually pleasure and surprise—of someone hearing a true high-end audio system for the first time underscores that high-end audio can be appreciated by everyone. If you enjoy music, you'll enjoy it more through a high-end system. It's that simple.

Finally, the goal of high-end audio is to make the equipment "disappear"; when that happens, we know that we have reached the highest state of communication between musician and listener. High-end audio isn't about equipment; it's about music.

The high-end credo holds that the less the musical signal is processed, the better. Any electronic circuit, wire, tone control, or switch degrades the signal—and thus the musical experience. This is why you won't find graphic equalizers, "spatial enhancers," "sub-harmonic synthesizers," or other such gimmicks in high-end equipment. These devices are not only departures from musical reality, they add unnecessary circuitry to the signal path. By minimizing the amount of electronics between you and the musicians, high-end audio products can maximize the directness of the musical experience. Less is more.

Imagine yourself standing at the edge of the Grand Canyon, feeling overwhelmed by its grandeur. You experience not only the vastness of this massive sculpture carved into the earth, but all its smaller features jump out at you as well, vivid and alive. You can discern fine gradations of hue in the rock layers—distinctions between the many shades of red are readily apparent. Fine details of the huge formations are easily resolved simply by your looking at them, thus deepening your appreciation. The contrasts of light and shadow highlight the apparently infinite maze of cracks and crevasses. The longer and closer you look, the more you see. The wealth of sensory input keeps you standing silently at the edge, in awe of nature's unfathomable beauty.

Now imagine yourself looking at the Grand Canyon through a window made of many thicknesses of glass, each one less than perfectly transparent. One pane has a slight grayish opacity that dulls the vivid hues and obliterates the subtle distinctions between similar shades of color. The fine granular structure of the next pane diminishes your ability to resolve features in the rock. Another pane reduces the contrast between light and shadow, turning the Canyon's immense depth and breadth into a flat canvas. Finally, the windowframe itself constricts your view, destroying the Canyon's overall impact. Instead of the direct and immediate reality of standing at the edge of the Grand Canyon, what you see is gray, murky, lifeless, and synthetic. You may as well be watching it on television.

Hearing reproduced music through a mediocre playback system is like looking at the Grand Canyon through those panes of glass. Each component in the playback chain—CD player, turntable, preamplifier, power amplifier, loudspeakers, and the cables that connect them—in some way distorts the signal passing through it. One product may add a coarse, grainy character to instrumental textures. Another may reduce the dynamic contrasts between loud and soft, muting the composer's or performer's expression. Yet another may cast a thick, murky pall over the music, destroying its subtle tonal colors and overlaying all instruments with an undifferentiated timbre. Finally, the windowframe—that is, the electronic and mechanical playback system—diminishes the expanse that is the musicians' artistic intent.

High-end audio is about removing as many panes of glass as possible, and making those that remain as transparent as they can be. The fewer the panes, and the less effect each has on the information passing through it, the closer we get to the live experience and the deeper our connection with the musical message.

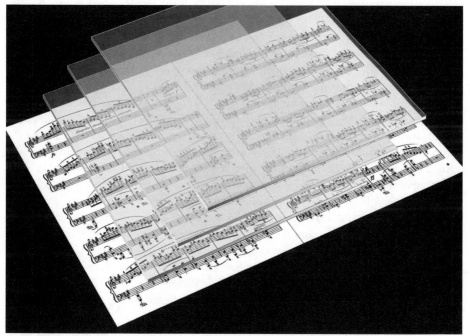

Each component in an audio system can be thought of as a piece of glass through which we experience music. (Courtesy AudioQuest)

Why are high-end audio products more transparent windows on the musical event than mass-market "stereo systems"? High-end products are designed to *sound* good—that is, like the real thing. They're not necessarily designed to perform "well" according to some arbitrary technical specification. The true high-end designer *listens* to the product during its development, changing parts and trying different techniques to produce the most realistic sound possible. He combines technical skill with musical sensitivity to create a product that best conveys the musical experience. This dedication often becomes a zealous pursuit, involving many hundreds of listening hours and painstaking attention to every factor that influences the sound. Often, a more expensive part will be included to improve the product's sound, while the retail price remains the same. The higher cost of this musically superior part comes off the company's bottom line. Why? Because the high-end designer cares deeply about music and its reproduction.

3

Chapter 1

Conversely, mass-market audio components are often designed to look good "on paper"—on the specification sheet—sometimes at the expense of sound quality. A good example of this is the "THD wars" of the 1970s and '80s. THD stands for Total Harmonic Distortion, a specification widely used by uneducated consumers as a measure of amplifier quality. (If you've done this, don't worry; before I learned more about audio, I, too, looked at THD figures.) The lower the THD, the better the amplifier was perceived to be. This led the electronics giants to produce products with vanishingly low THD numbers. It became a contest to see which brand had the most zeros after the decimal point in its THD specification (0.001%, for example). Many buyers bought receivers or amplifiers solely on the basis of this specification.

Although low THD is a worthy design goal, the problem arose in *how* those extremely low distortion figures were obtained. A technique to reduce distortion in amplifiers is called "feedback"—taking part of the output signal and feeding it back to the input. Large amounts of feedback reduce THD, but cause all kinds of other problems that degrade the amplifier's musical qualities. Did the electronics giants care that the large amounts of feedback induced to reduce their products' THD measurements actually made those products sound *worse*? Not a chance. The only thing that mattered was making a commodity that would sell in greater quantity. They traded musical performance for an insignificant technical specification that was sold to the public as being important. Those buyers choosing components on the basis of a specification sheet rather than listening ended up with poor-sounding systems. Ironically, the amplifiers that had the lowest THDs probably had the lowest quality of sound as well.

This example illustrates the vast difference between mass-market manufacturers' and high-end companies' conceptions of what an audio component should do. High-end manufacturers care more about how the product sounds than about how it performs on the test bench. They know that their audience of musically sensitive listeners will buy on the basis of sound quality, not specifications.

High-end products are not only designed by ear, but are often hand-built by skilled craftspeople who take pride in their work. The assemblers are often audiophiles themselves, building the products with as much care as if the products were to be installed in their own homes. This meticulous attention to detail results in a better quality of construction, or *build quality*. Better build quality can not only improve a product's sound, but increase its long-term reliability as well. Moreover, beautifully hand-crafted components can inspire a pride in ownership that the makers of mass-produced products can't hope to match.

High-end audio products are often backed by better customer service than mass-market products. Because high-end manufacturers care more about their products and customers, they generally offer longer warranties, more liberal exchange policies, and better service. It is not uncommon for high-end manufacturers to repair products out of warranty at no charge. This isn't to say you should expect such treatment, only that it sometimes happens with high-end and is unthinkable with mass-market products. High-end companies care about their customers.

What is High-End Audio?

These attributes also apply to high-end specialty retailers. The high-end dealer shares a passion for quality music reproduction and commitment to customer service. If you're used to buying audio components at a mass-market dealer, you'll be pleasantly surprised by a visit to a high-end store. Rather than trying to get you to buy something that may not be right for you, the responsible high-end dealer will strive to assemble a system that will provide the greatest long-term musical pleasure. Such a dealer will put your musical satisfaction ahead of this month's bottom line.

Finally, most high-end products are designed and built in America by American companies. In fact, American-made audio components are highly regarded throughout the world. More than 40% of all American high-end audio production is exported to foreign countries, particularly the Far East. This is true even though high-end products cost about twice as much abroad as they do in the U.S., owing to shipping, import duties, and importer profit. The enthusiasm for American high-end products abroad is even more remarkable when one remembers the popular American misperception that the best audio equipment is made in Japan.

On a deeper level, high-end products are fundamentally different from mass-market products, from their conception, purpose, design, construction, and marketing. In all these differences, what distinguishes a high-end from a mass-market product is the designer's caring attitude toward music. He isn't creating boxes to be sold like any other commodity; he's making musical instruments whose performance will affect how his customers experience music. The high-end component is a physical manifestation of a deeply felt concern about how well music is reproduced, and, by extension, how much it is enjoyed by the listener.

The high-end designer builds products he would want to listen to himself. Because he cares about music, it matters to him how an unknown listener, perhaps thousands of miles away, experiences the joy of music. The greater the listener's involvement in the music, the better the designer has done his job.

To the high-end designer, electronic or mechanical design isn't merely a technical undertaking—it's an act of love and devotion. Each aspect of a product's design, technical as well as musical, is examined in a way that would surprise those unaccustomed to such commitment. The ethos of music reproduction goes to the very core of the high-end designer's being; it's not a job he merely shows up for every day. The result is a much more powerful and intimate involvement in the music for the listener than is possible with products designed without this dedication.

What *is* high-end audio? What is high-end sound? It is when the playback system is forgotten, seemingly replaced by the performers in your listening room. It is when you feel the composer or performer speaking across time and space *to you*. It is feeling a physical rush during a musical climax. It is the ineffable roller-coaster ride of emotion the composer somehow managed to encode in a combination of sounds. It is when the physical world disappears, leaving only your consciousness and the music.

That is high-end audio.

2
Getting Started—Defining Your System

Before delving into the next chapters that describe in detail the specific components that make up an audio system, let's look at an overview of those components, how they might fit together, and options for putting together your own system.

At the very front of the playback chain are the *source components*. This category of product includes CD players, FM tuners, DVD players, satellite radio receivers (XM or Sirius), and turntables, to name a few. Any product that delivers a signal to the rest of your system is a source component.

The source components feed the next step in the playback chain, the control and amplification components. These components can be housed in separate chassis, called a *preamplifier* and a *power amplifier*, respectively. The preamplifier is the Grand Central Station of a hi-fi system. It takes signals from the various source components and controls which source signals (CD or radio, for example) you listen to, adjusts the volume, and performs a few other functions. A power amplifier, by contrast, is the workhorse of a hi-fi system, taking in the preamplifier's output and boosting it to a level that can drive the loudspeakers (Fig.2-1).

Many quality audio systems will use an *integrated amplifier*, a product that combines the functions of a preamplifier and power amplifier in one chassis. Integrated amplifiers are less expensive than a separate preamplifier and power amplifier, easier to operate, and take up less space in your home.

Another option is the *receiver*, a product that combines a preamplifier, power amplifier, and radio tuner in one chassis. Most receivers today can tune in AM and FM signals as well as those from XM or Sirius Satellite Radio subscription-based radio services. Satellite radio reception requires a separate antenna and purchase of a monthly subscription.

Receivers come in two flavors: the traditional *stereo receiver* that will power two loudspeakers for music listening, and the *audio/video receiver* (AVR) that powers a multichannel loudspeaker array for home theater (Fig.2-2). By far the more prevalent of the two is the AVR, which forms the heart of a system that will play music as well as movies. The AVR has either five or seven channels of amplification, surround-sound decoding, and the ability to control video as well as audio source components—DVD players, hard-disk video recorders such as TiVo, and satellite receivers such as DirecTV or Dish Network.

Another option for those who want their audio system to do double-duty on music and movies is the audio/video *controller* and *multichannel power amplifier*. The A/V controller is analogous to the stereo preamplifier just described, but it also has the ability to control video source components. A multichannel amplifier is simply a power amplifier with more than two channels of amplification, usually five or seven. The separate controller and multichannel power amplifier are (usually) higher-quality (but more expensive) alternatives to the AVR (Fig.2-3).

7

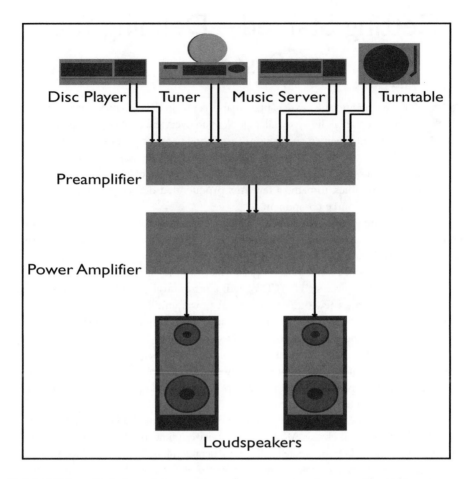

Fig.2-1 A high-quality 2-channel stereo system often uses a separate preamplifer and power amplifier.

Finally, those who want to listen to stereo and multichannel music but don't care about film soundtracks or home theater can choose a *multichannel preamplifier*. This component is identical to the stereo preamplifier we saw earlier, but has the ability to accommodate six audio channels rather than the stereo preamp's two channels.

Which of these system configurations you choose—preamplifier and power amplifier, integrated amplifier, stereo receiver, AVR, or controller and multichannel power amplifier—is influenced by whether you want an audio system for purely music listening or for playing back film soundtracks as part of a home-theater system. Your choice will also be influenced by your budget; a stereo receiver and two speakers is generally less expensive than an A/V controller, multichannel power amplifier, and five loudspeakers plus a subwoofer.

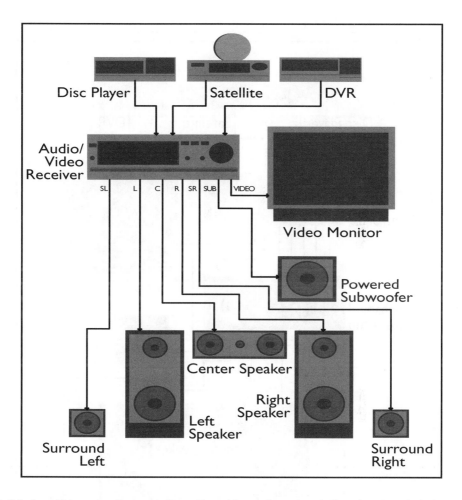

Fig.2-2 A multichannel audio system for music and home theater can be based on an audio/video receiver (AVR) and a 5.1-channel loudspeaker system.

To recap, here are your options, based on how you intend to use your audio system.

2-channel music listening only

Option A. Preamplifier and stereo power amplifier

Option B. Integrated amplifier

Option C. Stereo receiver

2-channel and multichannel music, as well as home theater

Option A. Audio/video receiver

Option B. Audio/video controller and multichannel power amplifier

9

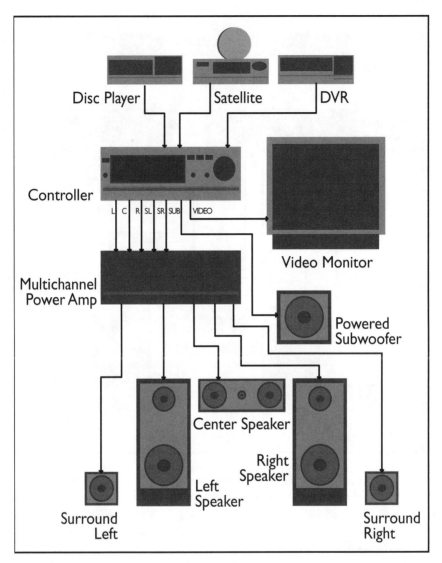

Fig.2-3 A multichannel audio system for music and home theater can be based on an a separate A/V controller and multichannel amplifier rather than on an A/V receiver.

The final link in the audio-system playback chain is the loudspeaker. The loudspeaker converts the high-power electrical output from the power amplifier into sound. If you choose a stereo preamplifier and stereo power amplifier, you'll need two loudspeakers, left and right. For reproducing multichannel music sources and film soundtracks, you'll select a multichannel speaker system consisting of *left* and *right* speakers, as well as a *center-channel* speaker and *surround* speakers. Home-theater systems often employ an additional speaker called a *subwoofer* for reproducing low bass sounds. The left, center, and right speakers are arrayed across the front of your room, with the left and right speakers on either side of your television and the center speaker positioned above or below the TV. The surround speakers are located to the sides of, or behind, the listening/viewing couch or chair.

Of these options, the one most likely to deliver the highest audio quality is the separate preamplifier and power amplifier (the reasons are discussed in Chapter 8). The next-best sounding choice is probably the integrated amplifier, followed by the stereo receiver. Although today's best audio/video receivers sound better than those of a few years ago, it's unlikely that an AVR will deliver the same quality of sound as a separate preamplifier and power amplifier, or even an integrated amplifier. Note, however, that these differences in sound quality are most apparent, and most important, for music listening, and less so when reproducing film soundtracks. If an AVR has sufficient output power (measured in the familiar "watts per channel") and is of reasonably good quality, it will deliver satisfying performance for the vast majority of home-theater enthusiasts. The differences lie in how these various components reproduce music; dedicated audio-only products will deliver a more engaging musical experience. (Chapter 4 describes the specific sonic qualities that affect how we perceive and connect with reproduced music.)

This last paragraph suggests that it's impossible to have a no-compromise system that performs equally well on music and home theater. There's yet one more system configuration option that combines the best of both worlds to deliver state-of-the-art music reproduction (if that's your goal) with the best possible reproduction of film soundtracks. That system employs both a stereo preamplifier as well as an audio/video controller. Two-channel music sources feed the preamplifier, which in turns drives the power amplifier, just as in a conventional 2-channel music-only system. But A/V sources (DVD player, satellite receiver, cable, digital video recorder such as TiVo) feed the controller, with the controller's left and right outputs driving one of the preamplifier's 2-channel inputs. The left and right channels of the film soundtrack or multichannel music source simply pass through the preamplifier to the power amplifier as though it weren't there. It goes without saying that this is the most expensive possible configuration, but it has the benefit of keeping critical 2-channel music signals out of the A/V controller. This system configuration, shown in Fig.2-4, is explained in more detail in Chapter 10.

Finally, there's one more path to getting great-sounding music and terrific home-theater: completely separate systems in different rooms of your home. Each can be optimized for the specific job, although this option is by far the most expensive, and also allows audio gear to take over another room in your home.

We'll look at each component of an audio system in Chapters 5–11, and how to assemble and fine-tune that system in Chapter 12. But first, let's look at how you go about choosing just the right combination of components for your system.

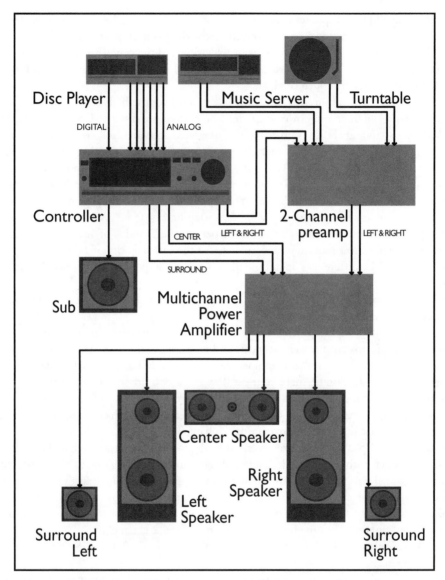

Fig.2-4 A no-compromise system for stereo music, multichannel music, and home-theater reproduction will use both a controller and a 2-channel preamplifier.

3

How to Choose an Audio System

Choosing a high-quality music-reproduction system is one of the more important purchasing decisions you'll make. Unlike buying home appliances, your selections in components will influence how deeply you appreciate and enjoy an art form—music. A great-sounding system can even change your lifestyle as music assumes a greater importance in your life. A hi-fi system is a vehicle for exploring the world of music; the better the system, the further and wider that vehicle will take you.

Although selecting hi-fi components may seem a daunting task, a little knowledge and preparation will go a long way toward realizing your dream system—and staying within your budget. The informed shopper knows that choosing the right components, matching those components to each other, and setting them up carefully are more important than having a big bank account. This chapter will teach you to become a wise shopper and show you the path to assembling the most musically and aesthetically satisfying system possible for your money.

Defining Your Needs

The first step in choosing an audio system is deciding whether you want the system to reproduce 2-channel music, multichannel music, film soundtracks (and other video-based sources), or all of the above. Consider the different system configurations presented in the previous chapter and choose the type of system best suited to your needs.

Keep in mind that spreading out a fixed budget among home-theater products and five loudspeakers plus a subwoofer will generally not produce as satisfying results for music as would the same budget dedicated to a 2-channel stereo system. Two good loudspeakers (or amplifier channels) are better than five mediocre ones. Nonetheless, if home theater is important to you, by all means choose a multichannel system that will deliver both music and film soundtracks.

This decision of whether to choose a music-only or music and home-theater system isn't an either-or proposition. You can selectively tailor the system for better music or home-theater performance just by component selection and budget allocation. For example, the music lover who occasionally watches movies would invest more in the left and right loudspeakers, and less in the system's other speakers—subwoofer, center and surrounds. This scenario is particularly appropriate for the movie lover who favors character- and dialog-driven films rather than action movies.

Conversely, if your primary interest is in home theater, you'll probably choose a loudspeaker system that equally balances the quality among all the channels. Later in this chapter will look at specific budget allocations for each path.

A related topic for home-theater fans is dividing the system budget between a video display and an audio system. The most common mistake made by the average home-theater shopper is to blow nearly all the budget on that big, bright, colorful plasma panel and find that there's very little money left for the audio system. Sound is fully half of the home-theater experience; don't make it an afterthought.

Once you've decided on the system's overall architecture, you need to match the equipment to your room and lifestyle. Just as a pickup truck is better suited to the farmer and a compact car to the city dweller, a hi-fi system ideal for a small New York City apartment would be entirely inadequate in a large suburban home. The hi-fi system must not only match your musical taste, as described in the next chapter, but must also suit your room and listening needs. (The following section is only an overview of how to choose the best system. More detailed information on how to select specific components is contained in Chapters 5–11.)

Many of the guidelines are fairly obvious. First, match the loudspeaker size to your listening room. Large, full-range loudspeakers don't work well in small rooms. Not only are large loudspeakers physically dominating, they tend to overload the room with bass energy. A loudspeaker that sounds fine in a 17' by 25' room will likely be thick, boomy, and bottom-heavy in a 12' by 15' room. The bass performance you paid dearly for (it's expensive to get correct deep-bass reproduction) will work against you if the loudspeaker is put in a small room. For the same money, you could buy a superb minimonitor whose build cost was put into making the upper bass, midrange, and treble superlative. You win both ways with the minimonitor: your room won't be overloaded by bass, and the minimonitor will likely have much better soundstaging and tonal purity. Conversely, a minimonitor just won't fill a large room with sound. The sense of power, dynamic drive, deep-bass extension, and feeling of physical impact so satisfying in some music just doesn't happen with minimonitors. If you've got the room and the budget, a full-range, floorstanding loudspeaker is the best choice.

Another important consideration is amplifier power. The amplifier should be matched to the loudspeakers and the room. A larger room requires more amplifier power, as does a loudspeaker of low sensitivity. These issues are described in detail in the power amplifier and loudspeaker chapters (8 and 9).

Setting Your Budget

How much you should spend on a music playback system depends on two things: your priorities in life and your financial means.

Let's take the priorities first. One person may consider a $2000 stereo system an extravagance, yet not bat an eye at blowing $7000 on a European vacation. Conversely, another person of similar means would find a $7000 vacation a waste of money when there is so much great hi-fi on the wish list. The first person probably considers music as merely a dispensable diversion while driving to work. To the second person, however, enjoying and appreciating music is a vital aspect of human existence. How much of your disposable income you should spend on a hi-fi system is a matter of how important music is to you, and only you can decide that. Owning a good hi-fi system will probably elevate music listening to a much higher priority in your life.

The second factor—your financial means—can often suggest an audio-system budget. One general guideline is to spend about 10% of your annual income on a system. If you can afford a high-quality system, I encourage you to invest in quality components. I'm often amazed to hear stories of well-heeled individuals who appreciate music yet own poor-quality audio systems.

Whatever your financial means, I strongly urge you to establish a significant budget for your high-end system. The expenditure may seem high at the time, but you will be rewarded night after night and year after year with your favorite music wonderfully reproduced. A year or two from now, as you enjoy your system, the money spent will have been forgotten, but the pleasure will continue—a good music-playback system is a lasting and fulfilling investment. Moreover, if you buy a quality system now, you won't want to sell it or trade it in for something better later. It is sometimes false economy to "save" money on a less than adequate system. Do it right the first time.

There's another way of determining how much to spend for a hi-fi system: Find the level of quality you're happy with and let that set your budget. Visit your dealer and have him play systems of various levels of quality. You may find yourself satisfied with a moderately priced system—or you may discover how good reproduced music can sound with the best equipment, and just *have* to have it.

The Complete vs. the Incremental Purchase

After you've established a budget, you must decide which of the following three ways of buying a hi-fi is best for you:

1) Buy an entire system made up of the finest components
2) Buy an entire system made up of components within a fixed budget
3) Buy just a few components now and add to the system as finances permit

The first option doesn't require much thought—just a large bank account. The purchaser of this system needn't worry about budgets, upgrading, and adding components later. Other elements of buying a high-end system (I'll talk about these later) do apply to the cost-no-object audiophile: allocating the budget to specific components, dealing with the retailer, and home auditioning. But this sort of listener isn't under the same financial constraints that force the tough choices inherent in the other two options. Just choose a top-notch system and start enjoying it.

Most of us, however, don't enjoy such luxury, and must live within the set budgets of options 2 or 3. Option 2 is to spend the entire budget on a complete system now. In option 3, you'll spend your entire budget on just a few, higher-quality components and add the other pieces later as money becomes available. We'll call option 2 the *complete purchase*, and option 3 the *incremental purchase*. There are advantages and disadvantages to each approach.

Buying an entire system at once means that the overall budget must be spread among all the components that make up an audio system. Consequently, you'll have less to spend on each component. You may not get the quality of components you'd hoped for, but the system will be complete and you can start enjoying it right away. This type of purchase is more suited to the music lover who doesn't want to think about equipment, but wants a system he can set up, forget, and use to enjoy music or home theater.

The listener more inclined to treat audio as a hobby, or who has her sights set on a more ambitious system, will buy just a few components now and add to the system as finances permit. This listener may spend the same amount of money as our first listener, but on just a pair of loudspeakers instead of on a whole system. She will use her old receiver or integrated amplifier, turntable, or CD player until she can afford the electronics and sources she really wants. Her system will be limited by the receiver's performance; she won't immediately get the benefit of her high-end loudspeakers. But when she *does* buy her dream electronics, she'll have a truly first-rate system. This approach requires more patience, commitment, and, in the long term, more money. But it is one way to end up with a superlative system.

The audiophile buying a system piece by piece can benefit from rapidly changing technology. She can start with components that don't change much over the years—power amplifiers, for example—and wait to buy those products likely to get better and cheaper over time, such as digital source components.

Another benefit of adding components to your system one at a time is the ability to audition components in that system before buying. Rather than putting together a whole system in a store showroom based on your own auditioning or a salesperson's recommendations, the incremental buyer can carefully audition components and choose the best musical match for the rest of the system. This is a big advantage when assembling a hi-fi best suited to your musical tastes. This piece-by-piece approach is more for the hobbyist, and demands a deeper level of commitment to audio. It also requires a great deal more patience.

In short, if you want to get a good system, forget about the hardware, and just enjoy music, buy the entire system now. Given the same initial expenditure, you won't end up with the same-quality system as if you'd bought pieces slowly, but you'll be spending less money in the long run and can have the benefit of high-end music playback immediately. Moreover, you can go about your life without thinking about what piece of audio hardware to buy next. The complete purchase is recommended for the music lover who isn't—and doesn't want to be—an audio hobbyist. Many music lovers taking this approach, however, find themselves upgrading their systems piece by piece after discovering the rewards of owning a high-end audio system.

Conversely, the music lover on a budget who plans to build an ambitious system, and who takes a more active role in audio equipment, will probably build a system gradually. This listener will more likely read product reviews in magazines, visit the dealer often, and take equipment home for evaluation. By doing so she'll become a better listener and a more critical audiophile, as well as develop a broader knowledge of audio equipment.

Whichever approach you take, the information in the rest of this chapter—selecting a system suited to your listening, allocating the budget to specific components, and dealing with the retailer—applies equally well.

Value vs. Luxury Components

High-end audio components run the gamut from utilitarian-looking boxes to lavish, gold-plated, cost-no-object shrines. The packaging doesn't always reflect the quality of the electronics inside, but rather the manufacturer's product philosophy. Some companies try to offer the best sound for the least money by putting excellent electronics in inexpensive chassis. These are the so-called "value" products. Manufacturers of "luxury" components may put the same level of electronics in a lavish chassis with a 1"-thick front panel, lacquer-filled engraving, expensive metalwork, and custom-machined input jacks or terminal posts.

A designer of *very* expensive electronics once told me that he could sell his products for *half* the price if he used a cheap chassis. He felt, however, that the level of design and execution in his electronics deserved no less than the ultimate in packaging.

Some buyers demand elegant appearance and a luxury feel; others merely want the best sound for the least money. To some music lovers, appearance and elegant packaging are secondary to sound quality; they don't care what it looks like so long as it sounds good. Conversely, some audiophiles are willing to pay for gorgeous cosmetics, battleship build-quality, and all the trimmings that make some products exude elegance and luxury. There's an undeniable pride of ownership that accompanies the finest-built audio components.

When choosing high-end components, match your needs with the manufacturer's product philosophy. That way, you won't waste money on thick faceplates you don't care about, or spend money on a product that doesn't do justice to your home decor.

Another type of manufacturer puts mediocre or even poor electronics in a fancy, eye-catching package. Their market is the less sophisticated buyer who chooses on the basis of appearance and status (or the company's past reputation) rather than sound quality. These so-called "boutique" brands are not high-end, however, and should be avoided.

Allocating Your Budget to Specific Components

There are no set rules for how much of your total budget you should spend on each component in your system. Allocating your budget between components depends greatly on which components you choose, and your overall audio philosophy. Mass-market mid-fi magazines have been telling their readers for years to spend most of a hi-fi budget on the loudspeakers because they ultimately produce the sound. This thinking also suggests that all amplifiers and CD players sound alike; why waste money on expensive amplifiers and digital sources?

Chapter 3

The high-end listener makes different assumptions about music reproduction. A fundamental tenet of high-end audio holds that if the signal isn't good at the beginning of the reproduction chain, nothing downstream can ever improve it. In fact, the signal will be degraded by *any* product it flows through. High-end audio equipment simply minimizes that degradation. If your CD player is bright, hard, and unmusical, the final sound will be bright, hard, and unmusical. Similarly, the total system's performance is limited by the resolution of the worst component in the signal path. You may have superb loudspeakers and an excellent turntable and cartridge, but they'll be wasted with a poor-quality preamp in the signal chain. Quality-matching between components is essential to getting the most sound for your budget.

As explained in Chapters 8 and 9, loudspeaker sensitivity (how loudly the speaker will play for a given amount of amplifier power) greatly affects how large an amplifier you need to achieve a satisfying volume. High-sensitivity speakers need very little amplifier power. And because amplifier power is costly, it follows that a system with high-sensitivity speakers can be driven by lower-cost amplification—*provided that the amplification is of high quality*. Fortunately, manufacturers have recently responded to the need for relatively inexpensive high-performance amplification by designing integrated amplifiers with outstanding sound quality, but with lower output power. By putting the preamplifier and power amplifier in one chassis and cutting back on power output, manufacturers can put their high-end circuits in lower-priced products. The trick is to find those bargain integrated amplifiers that deliver truly high-end sound, and mate them with loudspeakers that not only have the appropriate sensitivity, but are also a good musical match. This approach will get you the best sound for the least money. If, however, cost is secondary to sound quality—that is, if you're willing to spend more for an improvement in sound—then buy the best separate preamplifier and power amplifier you can find.

For the following exercise, I assembled an imaginary 2-channel system of the components I'd choose if my audio budget totaled $2000. This hypothetical system follows a traditional audiophile approach. Here are the costs per item:

Integrated amplifier	$600
CD player	$400
Loudspeakers	$850
Interconnects and cables	$150
Total	**$2000**

As you can see, loudspeakers consumed 42% of the budget, the digital source took up another 20%, and the integrated amplifier consumed 30%. The remaining 8% was spent on interconnects and cables. These numbers and percentages aren't cast in stone, but they're a good starting point in allocating your budget. If you wanted to include a turntable, tonearm, and cartridge, the budget for the other components would have to be reduced.

How to Choose an Audio System

The 42% figure for loudspeakers is flexible, although it's a reliable guide. As described in Chapter 9, many moderately priced loudspeakers outperform much more expensive models. Use Chapter 9's guidelines on choosing loudspeakers to get the most performance for your loudspeaker dollar.

I've heard systems at this price level that are absolutely stunning musically. When carefully chosen and set up, a $2000 high-end system can achieve the essence of what high-quality music reproduction is all about—communicating the musical message. I've even heard a whole system with a list price of $1200 that was musical and enjoyable. The point isn't how much you spend on a hi-fi, but how carefully you can choose components to make a satisfying system within your budget.

These guidelines are for a dedicated 2-channel music system, not a multichannel music and home-theater system. The different requirements of home theater suggest a somewhat different budget allocation. If most of your listening time with a multichannel system is devoted to music, the budget guidelines are very similar to those described earlier, but with a shift of money from electronics to loudspeakers. That is, you should spend a bit more of your total budget on speakers. That's because home theater and multichannel music require five or six loudspeakers, and spreading out a fixed amount of money over five speakers rather than two inevitably results in compromised performance. If music is your primary consideration, put most of your speaker budget into the left and right speakers, with less devoted to the subwoofer, center, and surround speakers. Those readers who spend more time watching movies should spend even less on electronics and source and more on speakers. That's because the center channel loudspeaker plays an important role in the home-theater experience (it reproduces nearly all the dialogue), and choosing a quality model is of paramount importance.

In my experience, the sonic differences among amplifiers, preamplifiers, and source components are much less pronounced when watching movies than when listening to music. So much of our attention is taken away from the aural experience by the overwhelming visual sensory input. I've used audio/video receivers in my system (during product reviews) and found many of them lacking when reproducing music, but nothing short of thrilling for film-soundtrack playback.

Let's look at the same hypothetical $2000 budget, but this time spread over a home-theater system. We'll use an audio/video receiver instead of a stereo integrated amplifier, a multichannel loudspeaker package rather than left and right speakers, and a DVD player rather than a CD player.

A/V receiver	$550
DVD player	$300
Loudspeakers	$1000
Interconnects and cables	$150
Total	**$2000**

The individual component allocations are similar to that of the 2-channel music system, but we've shifted some of the money from amplification and the digital source to the loudspeakers.

If your budget is considerably larger than $2000, the percentage allocations to amplification, sources, loudspeakers, and cables are quite similar to the example shown above.

Upgrading a Single Component

Many audiophiles gradually improve their systems by replacing one component at a time. The trick to getting the most improvement for the money is to replace the least good component in your system. A poor-sounding preamp won't let you hear how good your CD player is, for example. Conversely, a very clean and transparent preamp used with a grainy and hard digital source will let you hear only how grainy and hard the digital source is. The system should be of similar quality throughout. If there's a quality mismatch, however, it should be in favor of high-quality source components.

Determining which component to upgrade can be difficult. This is where a good high-end audio retailer's advice is invaluable—he can often pinpoint which component you should consider upgrading first. Another way is to borrow components from a friend and see how they sound in your system. Listen for which component makes the biggest improvement in the sound. Finally, you can get an idea of the relative quality of your components by carefully reading the high-end audio magazines, particularly when they recommend specific components.

In Chapter 1, I likened listening to music through a playback system to looking at the Grand Canyon through a series of panes of glass. Each pane distorts the image in a different way. The fewer and more transparent the panes are, the clearer the view, and the closer the connection to the direct experience.

Think of each component of a high-end audio system as one of those panes of glass. Some of the panes are relatively clear, while others tend to have an ugly coating that distorts the image. The pane closest to you is the loudspeaker; the next closest pane is the power amplifier; next comes the preamp; and the last pane is the signal source (CD, SACD, DVD-A player or turntable). Your view on the music—the system's overall transparency—is the sum of the panes. You may have a few very transparent panes, but the view is still clouded by the dirtiest, most colored panes. This idea is shown graphically on page 3.

The key to upgrading a hi-fi system is getting rid of those panes—those components—that most degrade the music performance, and replacing them with clearer, cleaner ones. This technique gives you the biggest improvement in sound quality for the money spent.

Conversely, putting a very transparent pane closest to you—the loudspeaker— only reveals in greater detail what's wrong with the power amplifier, preamplifier, and source components. A high-resolution loudspeaker at the end of a mediocre electronics chain can actually sound worse than the same system with a lower-quality loudspeaker.

Following this logic, we can see that a hi-fi system can never be any better than its source components. If the first pane of glass—the source component—is ugly, colored, and distorts the image, the result will be an ugly, colored, and distorted view.

As you upgrade your system, you can start to see that other panes you thought were transparent actually have some flaws you couldn't detect before. The next upgrade step is to identify and replace what is *now* the weakest component in the system. This can easily become an ongoing process.

Unfortunately, as the level of quality of your playback system rises, your standard of what constitutes good performance rises with it. You may become ever more critical, upgrading component after component in the search for musical satisfaction. This pursuit can become an addiction and ultimately *diminish* your ability to enjoy music. Don't fall into that trap of caring so much about sound quality that you forget the reason you pursue high-quality audio equipment—connecting with the music.

How to Read Magazine Reviews

I've deliberately avoided recommending specific products in this book. By the time you would have read the recommendations, the products would likely have been updated or discontinued. The best source for advice on components is product reviews in high-end audio magazines. Because most of these magazines are published monthly, they can stay on top of new products and offer up-to-date buying advice. Many magazine reviewers are highly skilled listeners, technically competent and uncompromising in their willingness to report truthfully about audio products. There's a saying in journalism that epitomizes the ethic of many high-end product reviewers: "without fear or favor." The magazine should neither fear the manufacturer when publishing a negative review, nor expect favor for publishing a positive one. Instead, the competent review provides unbiased and educated opinion about the sound, build-quality, and value of individual products. Because high-end reviewers hear lots of products under good conditions, they are in an ideal position to assess the relative merits and drawbacks and report their informed opinions. The best reviewers have a combination of good ears, honesty, and technical competence.

You'll notice a big difference between reviews in high-end magazines and reviews in the so-called "mainstream slicks." The mass-market, mainstream magazines are advertiser-driven; their constituents are their advertisers. Conversely, high-end magazines are usually reader-driven; the magazines' goal is to serve their readers, not their advertisers. Consequently, high-end magazines often publish negative reviews, while mass-market magazines generally do not. Moreover, high-end magazines are much more discriminating about what makes a good component that can be recommended to readers. Mass-market magazines cater to the average person in the street, who, they believe (mistakenly, in my view), doesn't care about aspects of music reproduction important to the audiophile. The high-end product review is not only more honest, but much more discriminating in determining what is a worthy product. If you're reading a hi-fi magazine that never criticizes products, beware. Not all audio components are worth buying; therefore not all magazine reviews should conclude with a recommendation. Also beware of so-called "consumer" publications that regard an audio system as an appliance, not as a vehicle for communicating an art form. In their view, the improvements in performance gained by spending more are not worth the money, so they recommend the least expensive products with the most features. They take a similar approach to reviewing cars; why spend lots of money on a depreciating asset when all cars perform an identical function—moving you from point A to point B? It is simply not within their ethos to value spirited acceleration, tight handling, a luxurious interior, or the overall feel of the driving experience.

As someone who makes his living reviewing high-end audio products, I'll let you in on a few secrets that can help you use magazine reviews to your advantage. First, associate the review you're reading with the reviewer. Before reading the review, look at the byline and keep in mind who's writing the review. This way, you'll quickly learn different reviewers' tastes in music and equipment. Seek the guidance of reviewers with musical sensibilities similar to your own.

Second, don't assign equal weight to all reviews or reviewers. Audio reviewing is like any other field of expertise—there are many different levels of competence. Some reviewers have practiced their craft for decades, while others are newcomers who lack the seasoned veteran's commitment to the profession. Consider the reviewer's reputation, experience, and track record when giving a review credence. Also consider the reviewer's standards for what makes a good product.

Finally, listen to the products yourself. If two components are compared in a review, compare those products to hear if your perceptions match the reviewer's. Even if you aren't in the market for the product under review, listening to the components will sharpen your skills and put the reviewer's value judgments in perspective.

A common mistake among audiophiles looking for guidance is to select a component on the basis of a rave review without fully auditioning the product for themselves. The review should be a starting point, not a final judgment of component quality. It's much easier to buy a product because a particular reviewer liked it than it is to research the product's merits and shortcomings for yourself. Buying products solely on the basis of a review is fraught with danger. Never forget that a review is nothing more than one person's opinion, however informed and educated that opinion might be. Moreover, if the reviewer's tastes differ from yours, you may end up with a component you don't like. We all have different priorities in judging reproduced sound quality; what the reviewer values most—soundstaging, for example—may be lower in your sonic hierarchy. Your priorities are the most important consideration when choosing music-playback components. Trust your own ears.

In sum, reviews can be very useful, provided you:

• Get to know individual reviewers' sonic and musical priorities. Find a reviewer on your sonic and musical wavelength, and then trust his or her opinions.

• Compare your impressions of products to the reviewer's impressions. This will not only give you a feel for the reviewer's tastes and skill, but the exercise will make you a better listener.

• Don't assign the same weight to all reviews. Consider the reviewer's reputation and experience in the field. How many similar products has the reviewer auditioned? If a reviewer has heard virtually every serious CD player, for example, his or her opinion will be worth more than that of the reviewer who has heard only a few models.

• Don't buy—or summarily reject—products solely on the basis of a review. Use product recommendations as a starting point for your own auditioning. Listen to the product yourself to decide if that product is for you. Let *your* ears decide.

System Matching

It is a truism of high-end audio that an inexpensive system can often outperform a more costly and ambitious rig. I've heard modest systems costing, say, $1500 that are more musically involving than $50,000 behemoths. Why?

Part of the answer is that some well-designed budget components sound better than ill-conceived or poorly executed esoteric products. But the most important factor in a playback system's musicality is *system matching*. System matching is the art of putting together components that complement each other sonically so that the overall result is a musicality beyond what each of the components could achieve if combined with less compatible products. The concept of synergy—that the whole is greater than the sum of the parts—is very important in creating the best-sounding system for the least money.

System matching is the last step in choosing an audio system. You should have first defined the system in terms of your individual needs, set your budget, and established a relationship with a local specialty audio retailer. After you've narrowed down your choices, which products you select will greatly depend on system matching.

Knowing what components work best with other components is best learned by listening to a wide range of equipment. Many of you don't have the time—or access to many diverse components—to find out for yourselves what equipment works best with other equipment. Consequently, you must rely on experts for general guidance, and on your own ears for choosing specific equipment combinations.

The two best sources for this information are magazine reviews and your high-end audio dealer. Your dealer will have the greatest knowledge about products he carries, and can make system-matching recommendations based on his experience in assembling systems for his customers. Your dealer will likely have auditioned the products he sells in a variety of configurations; you can benefit from his experience by following his system-matching recommendations.

The other source of system-matching tips is magazine reviews. Product reviews published in reputable magazines will often name the associated equipment used in evaluating the product under review. The reviewer will sometimes describe his or her experiences with other equipment not directly part of the review. For example, a loudspeaker review may include a report on how the loudspeaker sounded when driven by three or four different power amplifiers. The sonic characteristics of each combination will be described, giving the reader an insight into which amplifier was the best match for that loudspeaker. More important, however, the sonic descriptions and value judgments expressed can suggest the *type* of amplifier best suited to that loudspeaker. By *type* I mean both technical performance (tubed vs. transistor, power output, output impedance, etc.) and general sonic characteristics (forward presentation, well-controlled bass, etc.).

These reports of system matching can extend beyond the specific products reported on in the review. A fairly good idea of which type of sonic and technical performance is best suited to a particular product can be gained from a careful reading of product reviews. For example, you may conclude that a particular loudspeaker needs to be driven by a large, high-current amplifier. This knowledge can then point you in the right direction for equipment to audition yourself; you can rule out low-powered designs. You can also get a feel for how professional reviewers assemble systems in the annual Recommended Systems feature in *The Absolute Sound*.

By reading magazine reviews, following your dealer's advice, and listening to combinations of products for yourself, you can assemble a well-matched system that squeezes the highest musical performance from your hi-fi budget.

Do's and Don'ts of Selecting Components

Some audiophiles are tempted to buy certain products for the wrong reasons. For example, many high-end products are marketed on the basis of some technical aspect of their design. A power amplifier may, for example, be touted as having "over 200,000 microfarads (µF) of filter capacitance," "32 high-current output devices," and a "discrete JFET input stage." While these may be laudable attributes, they don't guarantee that the amplifier will produce good sound. Don't be swayed by technical claims—listen to the product for yourself.

Just as you shouldn't make a purchasing decision based on specifications, neither should you base your decision solely on brand name. Many high-end manufacturers with solid reputations sometimes produce mediocre-sounding products. A high-end marquee doesn't necessarily mean high-end sound. Again, let your ears be your guide. I sometimes discover moderately priced products that sound as good—or very nearly as good—as products costing two or three times as much.

You should, however, consider the company's longevity, reputation for build quality, customer service record, and product reliability when choosing components. High-end manufacturers run the gamut from one-man garage operations to companies with hundreds of employees and advanced design and manufacturing facilities. The garage operation may produce good-sounding products, but may not be in business next year. This not only makes it hard to get service, but also greatly lowers the product's resale value.

High-end manufacturers also have very different policies regarding service. Some repair their products grudgingly, and/or charge high fees for fixing products out of warranty. Others bend over backward to keep their valued customers happy. In fact, some high-end audio companies go to extraordinary lengths to please their customers. One amplifier manufacturer who received an out-of-warranty product for repair not only fixed the amplifier free of charge, but replaced the customer's scratched faceplate at no cost! It pays in the long run to do business with manufacturers who have reputations for good customer service.

Another factor to consider before laying down your hard-earned cash is how long the product has been on the market. Without warning, manufacturers often discontinue products and replace them with new ones, or update a product to "Mark II" status. When this happens, the value of the older product drops immediately. If you know an excellent product is about to be discontinued, you can often buy the floor sample at a discount. This is a good way of saving money, provided the discount is significant. You end up with a lower price, plus all the service and support inherent in buying from an authorized and reputable dealer rather than from a private party.

The best source of advance information on new products and what's about to be discontinued are reports in audio magazines and websites (www.AVguide.com) from the annual Consumer Electronics Show.

Your Relationship with the Retailer

If, in the past, you've bought audio equipment only from mass-market retailers, you should expect to have an entirely different sort of relationship with a high-end dealer. The good specialty audio retailer doesn't just "move" boxes of electronics; he provides you with the satisfaction of great-sounding music in your home. More than just an equipment dealer, he's usually a dedicated audio and music enthusiast himself—he *knows* his products, and is often the best person to advise you on selecting equipment and system setup.

Consider the very different relationships between seller and buyer in the following scenarios. In the first, a used-car dealer in downtown Los Angeles is trying to sell a car to someone from out of town. The seller has only one shot at the buyer, and he intends to make the most of it. He doesn't care about return business, the customer's long-term satisfaction with the purchase, or what the customer will tell his friends about the dealer. It will be an adversarial relationship from start to finish.

Then consider a new-car dealer in Great Falls, Montana, selling a car to another Great Falls resident. For this dealer, return business is vital to his survival. So is customer satisfaction, quality service, providing expert advice on models and options, finding exactly the right car for the particular buyer, and giving the customer an occasional ride to work when he drops off his car for service. He knows his customers by name, and has developed mutually beneficial, long-term relationships with them.

If buying a mass-market hi-fi system is like negotiating with a used-car dealer in downtown L.A., selecting a high-end music system should be made within a relationship similar to the one enjoyed between the Montana dealer and his customers.

Take the time to establish a relationship with your local dealer. Make friends with him—it'll pay off in the long run. Get to know a particular salesperson and, if possible, the store's owner. Tell them your musical tastes, needs, lifestyle, and budget—then let them offer equipment suggestions. They know their products best, and can offer specific component recommendations. The good stores will regard you as a valued, long-term customer, not someone with whom they have one shot at making a sale. Don't shop just for equipment—shop for the retailer with the greatest honesty and competence.

Keep in mind, however, that dealers will naturally favor the brands they carry. Be suspicious of dealers who badmouth competing brands that have earned good reputations in the high-end audio press. The best starting point in assembling your system is a healthy mix of your dealer's recommendations and unbiased, competent magazine reviews.

Unfortunately, not all high-end dealers subscribe to the idea of serving their customers' long-term satisfaction. A few even take an elitist attitude toward their customers. If you encounter such a dealer, take your business elsewhere. The retailer is a key ingredient in realizing high-quality music reproduction in your home; don't settle for one that isn't committed to your satisfaction.

The high-end retailing business is very different from the mass-market merchandising of the low-quality "home entertainment" products sold in appliance emporiums. The specialty retailer's annual turnover is vastly lower than that of the mid-fi store down the street. Consequently, the specialty retailer's profit margin must be larger for him to stay in business. Don't expect him to offer huge discounts and price cuts on equipment to take a sale from a mid-fi store. Because the high-end dealer offers so much more than just pushing a box over the counter, his prices just can't be competitive. Instead, you should be prepared to pay full list price—or very close to it.

Here's why. After paying his employees, rent, lights, heat, insurance, advertising, and a host of other expenses, the specialty audio retailer can expect to put in his pocket about five cents out of every dollar spent in his store. Now, if he discounts his price by even as little as 5%, he is essentially working for free, and only keeping his doors open a little longer. If the dealer offers a discount or marks down demonstration or discontinued units, you should take advantage of these opportunities. But don't expect the dealer to automatically discount; he deserves the full margin provided by the product's suggested retail price.

In return for paying full price, however, you should receive a level of service and professionalism second to none. Expect the best from your dealer. Spend as much time as you feel is necessary auditioning components in the showroom before you buy. Listen to components at home in your own system before you buy. Ask the retailer to set up your system for you. Exploit the dealer's wide knowledge of what components are best for the money. Use his knowledge of system matching to get the best sound possible on a given budget. And if one of your components needs repair, don't be afraid to ask for a loaner until yours is fixed. The dealer should bend over backward to accommodate your needs.

If you give the high-end dealer your loyalty, you can expect this red-carpet treatment. This relationship can be undermined, however, if, to save a few dollars, you buy from a competitor a product that your dealer also sells. If the product purchased elsewhere doesn't sound good in your system, don't expect your local dealer to help you out. Further, don't abuse the home audition privilege. Take home only those products you're seriously considering buying. If the dealer let everyone take equipment home for an audition, he'd have nothing in the store to demonstrate. The home audition should be used to confirm that you've selected the right component through store auditioning, magazine reviews, and the dealer's recommendations. The higher price charged by the dealer may seem hard to justify at first, but in the long run you'll benefit from his expertise and commitment to you as a customer.

If you don't live close to any high-end dealers, there are several very good mail-order companies that offer excellent audio advice over the phone. They provide as much service as possible by phone, including money-back guarantees, product exchanges, and component-matching suggestions. You can't audition components in a store, but you can often listen to them in your system and get a refund if the product isn't what you'd hoped it would be.

In short, if you treat your dealer right, you can expect his full expertise and commitment to getting you the best sound possible. There's absolutely no substitute for a skilled dealer's services and commitment to your satisfaction.

That's why your dealer's skill and knowledge are crucial to realizing the best possible sound for your budget.

Used Equipment

A used audio component often sells for half its original list price, making used gear a tempting alternative. The lower prices on used high-end components provide an opportunity to get a high-quality system for the same budget as a less ambitious new system. Moreover, buying used gear lets you audition many components at length. If you find a product you like, keep it. If you don't like the sound of your used purchase, you can often sell it for no less than what you paid.

There are two ways of buying used equipment: from a dealer, and from a private party. A retailer may charge a little more for used products, but often offers a short warranty (60 or 90 days), and sometimes exchange privileges. Buying used gear from a reputable retailer is a lot less risky than dealing with a private party (unless you're buying from a friend).

The audiophile inclined to buy used equipment can often get great deals. Some audiophiles simply *must* own the latest and greatest product, no matter the cost. They'll buy a state-of-the-art component one year, only to sell it the next to acquire the current top of the line. These audiophiles generally take good care of the equipment and sell it at bargain-basement prices. If you can find such a person, have him put your name at the top of his calling list when he's ready to sell. You can end up with a superb system for a fraction of its original selling price.

A few pitfalls await the buyer of used equipment, though the disadvantages listed below apply mostly to buying from a private party rather than from a reputable dealer. First, there are no assurances that the component is working properly; the product could have a defect not apparent from a cursory examination. Second, the used product could be so outdated that its performance falls far short of new gear selling for the same price—or less—than the used component. This is especially true of CD players, CD transports, and digital processors. Third, a used product carries no warranty; you'll have to pay for any repair work. Finally, you must ask why the person is selling the used equipment. All too often, a music lover doesn't do his homework and buys a product that doesn't satisfy musically or work well with the rest of his system. If you see lots of people selling the same product, beware—it's a sign that the product has a fundamental musical flaw. Finally, buying used equipment from a private party eliminates everything that makes your local specialty retailer such an asset when selecting equipment: You don't get the dealer's expert opinions, home audition, trade or upgrade policy, dealer setup, warranty, dealer service, loaner units, or any of the other benefits you get from buying new at a dealer.

Approach used components with caution; they can be a windfall—or a nightmare.

Product Upgrades

Many manufacturers improve their products and offer existing customers the option of upgrading their components to current performance. This is the "Mark II" (or III or IV) designation on some products. The dealer can usually handle sending your component back to the factory. Some manufacturers prefer to deal with the customer directly, saving the dealer markup and keeping the upgrade price lower.

I have ambivalent feelings about manufacturer upgrades. Some upgrade programs provide lasting value to a company's customers. Enjoying the benefits of a company's advances without selling a component at a loss and buying a new one can be a wonderful bonus. Conversely, some manufacturers think of an upgrade program as a profit center, charging large amounts for even minor improvements. Consider a company's track record when evaluating potential future upgrades. For products that are upgradeable via software updates, ask how much these revisions will cost. Some manufacturers charge nothing, or a nominal fee, for updated software.

———

If you read the rest of this book, subscribe to one or more reputable high-end magazines, and follow these guidelines, you'll be well on your way to making the best purchasing decisions—and having high-quality music reproduction in your home.

One last piece of advice: After you get your system set up, forget about the hardware. It's time to start enjoying the music.

4

Becoming a Better Listener

Critical listening—the practice of evaluating the quality of audio equipment by careful analytical listening—is very different from listening for pleasure. The goal isn't to enjoy the musical experience, but to determine if a system or component sounds good or bad, and what *specific characteristics* of the sound make it good or bad. You want to critically examine what you're hearing so that you can form judgments about the reproduced sound. You can then use this information to evaluate and choose components, and to fine-tune a system.

Listening vs. Measurement

Evaluating audio equipment by ear is essential—today's technical measurements simply aren't advanced enough to characterize the musical performance of audio products. The human hearing mechanism is vastly more sensitive and complex than the most sophisticated test equipment now available. Though technical performance is a valid consideration when choosing equipment, the ear should always be the final arbiter of good sound. Moreover, the musical significance of sonic differences between components can only be judged subjectively.

Good technical performance can contribute to high-quality musical performance, but it doesn't tell you what you really want to know: how well the product communicates the musical message. To find that out, you must listen. I have auditioned hundreds of audio products for review and measured their technical performance in a test laboratory. My experience overwhelmingly indicates that much more about the quality of an audio component can be learned in the listening room than in the test lab.

Many newcomers to high-performance music reproduction—and even a small fringe group of experienced audiophiles—question the need for listening to evaluate products. They believe that measurements can tell them everything they need to know about a product's performance. And since these measurements are purely "objective," why interject human subjectivity through critical listening?

The answer is that the common measurements in use today were created decades ago as design tools, not as descriptors of sound quality. The test data generated by a typical mix of audio measurements were never meant to be a representation of musical reality, only a rough guide when designing. For example, an amplifier circuit that had 1% harmonic distortion was probably better than one with 10% harmonic distortion. It doesn't follow that a harmonic-distortion specification in any way describes the sound of that amplifier.

A second problem is that audio test-bench measurements attempt to quantify a variety of two-dimensional phenomena: how much distortion the product introduces, its frequency response, noise level, and other factors. But music listening is a *three-*

dimensional experience that is much more complex than any set of numbers can hope to quantify. How can you reduce to a series of mathematical symbols the ability of one power amplifier, and not another, to make the hair on your arms stand up? Or the feeling that a vocalist is singing directly *to you*?

No matter how many measurements are gathered about the product's technical performance, they still don't tell you how well that product communicates the music. If I had to choose between two unknown CD players as my main source of music for the next five years, I'd rather have ten minutes with each player in the listening room than ten hours with each in the test lab. Today's measurements are crude tools that are inferior to the most powerful test instrument ever devised: the human brain.

Introduction to Critical Listening

Knowing what sounds good and what doesn't is easy; most people can tell the difference between excellent and poor sound. But discovering *why* a product is musically satisfying or not, and the ability to recognize and describe subtle differences in sound quality, are learned skills. Like all skills, that of critical listening improves with practice: The more you listen, the better a listener you'll become. As your ear improves, you'll be able to distinguish smaller and smaller differences in reproduced sound quality—and be able to describe *how* two presentations are different, and why one is better.

This chapter defines the language of critical listening, describes what to listen for, and outlines the procedures for setting up valid listening comparisons. It will either get you started in critical listening, or help you become a more highly skilled listener.

Audiophile Values

A general discussion of audiophile values is important in understanding the next sections of this chapter. Here are some broad statements about what distinguishes good from superlative sound quality, and audiophile values in reproduced music.

Good sound is only a means to the end of musical satisfaction; it is not the end itself. If a neighbor or colleague invites you over to hear his hi-fi system, you can tell immediately whether he's a music lover or a "hi-fi buff" more interested in sound than in music. If he plays the music very loud, then turns it down after 30 seconds to seek your opinion (approval), he's probably not a music lover. If, however, he sits you down, asks what kind of music *you* like, plays it at a reasonable volume, and says or does nothing for the next 20 minutes while you both listen, it's likely that this person holds audiophile values or simply cares a lot about music.

In the first example, the acquaintance tried to impress you with *sound*. In the second case, your friend also wanted to impress you with his system, but by its ability to express the *music*, not shake the walls. This is the fundamental difference between "hi-fi enthusiasts" and music lovers. (You can use the same test to immediately tell what kind of hi-fi store you're in. If anyone pulls out a CD of trains, sonic booms, Shuttle launches, or jet takeoffs, run for cover.)

All audio components affect the signal passing through them. Some products add artifacts (distortion) such as a *grainy treble* or a *lumpy bass*. Others subtract parts of the signal—for example, a loudspeaker that doesn't go very low in the bass. (Listening terms are defined later in this chapter.) A fundamental audiophile value holds that sins of commission (adding something to the music) are far worse than sins of omission (removing something from the music). If parts of the music are missing, the ear/brain system subconsciously fills in what isn't there; you can still enjoy listening. But if the playback system adds an artificial character to the sound, you are constantly reminded that you're hearing a reproduction and not the real thing.

Let's illustrate this sins-of-commission/omission dichotomy with two loud-speakers. The first loudspeaker—a three-way system with a 15" woofer in a very large cabinet—sells for a moderate price in a mass-market appliance store; it plays loudly and develops lots of bass. The second loudspeaker sells for about the same price, but is a small two-way system with a 6" woofer. It doesn't play nearly as loudly, and produces much less bass. While you need a refrigerator dolly to move the first loudspeaker, you can almost hold the second loudspeaker in your outstretched hand.

The behemoth loudspeaker has some problems: The bass is boomy, thick, and overwhelming. All the bass notes seem to have the same pitch. The very prominent treble is coarse and grainy, and the midrange has a big peak of excess energy that makes singers sound as if they have colds.

The small loudspeaker has no such problems. The treble is smooth and clear, and the midrange is pure and open. It has, however, very little bass by comparison, won't play very loudly, and doesn't produce a physical sensation of sound hitting your body.

The first loudspeaker commits sins of commission, *adding* unnatural artifacts to the sound. The bass peaks that make it sound boomy, the grain overlaying the treble, and the midrange colorations are all additive distortions.

The second loudspeaker's faults, however, are of omission. It *removes* certain elements of the music—low bass and loud peaks—but leaves the remainder of the music intact. It doesn't add grain to the treble, thickness to the bass, or colorations to the midrange.

There's no doubt that the second loudspeaker will be more musically satisfying. The first loudspeaker's additive distortions are not only much more musically objectionable, they also constantly remind you that you are listening to artificially reproduced music. The second loudspeaker's flaws are of a nature that allows you to forget that you're listening to loudspeakers. In the reproduction of music, addition is far worse than subtraction.

Another audiophile value holds that even small differences in the quality of the musical presentation are important. Because music matters to us, we get excited by *any* improvement in sound quality. Moreover, there isn't a linear relationship between the magnitude of a sonic difference and its musical significance. A quality difference can be sonically small but musically large.

While reviewing a revelatory new state-of-the-art digital-to-analog converter, I listened to a piece of music I'd heard hundreds of times before. The piece, performed by a five-member group, had vocals and very long instrumental breaks. During the instrumental breaks, the vocalist played percussion instruments. Through lesser-quality digital processors, the percussion had always been just another sound fused into the music's tapestry; I'd never heard it as a separate instrument played by the vocalist. The group seemed to become a four-piece ensemble when the vocalist wasn't singing; I never heard the percussion as separate from the rest of the music.

The new digital processor was particularly good at resolving individual instruments and presenting them not as just more sounds homogenized into the overall musical fabric, but as distinct entities. Consequently, when the instrumental break came, I heard the percussion as a separate, more prominent instrument. In my mind's eye, and for the first time, the vocalist never left—she remained "on stage," playing the percussion instruments. By just this "small" change in the presentation, the band went from being a quartet to a quintet during the instrumental breaks. The "objective" difference in the electrical signal must have been minuscule; the subjective musical consequences were profound.

This is why small differences in the musical presentation are important—*if* you care deeply enough about music and about how well it is reproduced. "Small" improvements can have large subjective consequences. This example highlights the inability of measurements alone to characterize audio equipment performance. Measurements on the digital-to-analog converter in question indicated no technical attributes that would have contributed to my perceptions. More fundamentally, how can a number representing some aspect of the digital converter's technical performance begin to describe the musical significance of the change I heard?

Much of music's expression and meaning can be found in such minutiae of detail, subtlety, and nuance. When such subtlety is conveyed by the playback system, you feel a vastly deeper communication with the musicians. Their intent and expression are more vivid, allowing you to more deeply appreciate their artistry. For example, if you compare two performances of Max Reger's Sonata in D Major for solo violin—one competent, the other superlative—you could say that, on an objective basis, they were virtually identical. Both performers played the same notes at about the same tempo. The difference in expression is in the nuances—the inspired subtleties of rubato, tempo, emphasis, articulation, and dynamics that bring the performance to life and convey the piece's musical meaning and intent. This example is analogous to the difference between mediocre and superb music playback systems, and why small differences in sound quality can matter so much. High-end audio is about reproducing these nuances so that you can come one step closer to the musical expression.

The sad but universal truth about audio equipment is that, any time you put a signal into an audio component, it *never* comes out better at the other end. You therefore want to keep the signal path as simple as possible, to remove any unnecessary electronics from between you and the music. This is why inserting equalizers and other such "enhancers" into the signal chain is usually a bad idea—the less done to the signal, the better. The advent of digital technology, however, has made possible some beneficial signal processing. (An example is digital room correction, described in Chapter 12.)

Pitfalls of Becoming a Critical Listener

There are dangers inherent in developing critical listening skills. The first is an inability to distinguish between critical listening and listening for pleasure. Once started on the path of critiquing sound quality, it's all too easy to forget that the reason you're involved in audio is because you love music, and to start thinking that every time you hear music, you must have an opinion about what's right and what's wrong with the sound. This is the surest path toward a condition humorously known as *Audiophilia nervosa*. Symptoms of *Audiophilia nervosa* include constantly changing equipment, playing only one track of a CD or LP at a time instead of the whole record, changing cables for certain music, refusing to listen to great music if it happens to be poorly recorded, and in general "listening to the hardware" instead of to the music.

But high-end audio is about making the hardware disappear. When listening for pleasure—which should be the vast majority of your listening time—forget about the system. Forget about critical listening. Shift into critical-listening mode only when you need to make a judgment, or just for practice to become a better listener. Draw the line between critical listening and listening for pleasure—and know when to cross it and when *not* to cross it.

There is also the related danger that your standards of sound quality will rise to such a height that you can't enjoy music unless it's "perfectly" reproduced—in other words, to the point that you can't enjoy music, period. Although it's not very high-quality reproduction, I get a great deal of pleasure from my car stereo and iPod—don't let being an audiophile interfere with *your* enjoyment of music, anytime, anywhere. When you can't control the sound quality, lower your expectations.

Sonic Descriptions and their Meanings

The biggest problem in critical listening is finding words to express our perceptions and experiences. We hear things in reproduced music that are difficult to identify and put into words. A listening vocabulary is essential not only to conveying to others what we hear, but also to recognizing and understanding our own perceptions. If you can attach a descriptive name to a perception, you can more easily recognize that perception when you experience it again.

By describing in detail the specific sonic characteristics of how electronic components change the sound of music passing through them, I hope to attune you to recognizing those same characteristics when you listen. After reading this next section, listen to two products for yourself and try to hear what I'm describing. It can be any two products—if you have a portable CD player, hook it up to your system and compare it to your home CD player. Even comparing a CD and an MP3 file made from that CD will get you started. The important thing is to start listening analytically. If you don't hear the sonic differences immediately, keep listening. The more you listen, the more sensitive you'll become to those differences.

tice this first-hand when I occasionally spend time listening critically in my
n with visiting manufacturers and designers of high-end equipment—
many of them highly skilled listeners. While we share many commonalities in determining what sounds good, there is a wide range of perception about what aspects of
the presentation are most important.

You should also know that recordings made with audiophile techniques are
more revealing of some aspects of reproduced sound than recordings made for mass
consumption. For example, a recording of classical music made in a concert hall with
very few microphones, a simple signal path, and high-quality recording equipment will
likely reveal more about a component's soundstaging performance than a pop recording made in a studio. Similarly, most mass-market recordings have almost no dynamic
range so that they sound "good" on a 4" car-stereo speaker. For these reasons, some of
the sonic terms described in this chapter apply much more to audiophile-quality
recordings than to mass-market ones.

It's also useful to understand the broad terms that describe the audio frequency band. The range of human hearing, which spans ten octaves from about 16Hz
(cycles per second) to 20,000Hz, or 20 kilohertz (20kHz), can be divided into the specific regions described below. Note that these divisions are somewhat arbitrary; you
can't say specifically that the lower treble begins at 2000Hz and not 2500Hz, for example. The table nonetheless provides a rough guideline for understanding the relationship between frequency ranges and their descriptive names.

Frequency Range

Lower Limit	Upper Limit	Description
16	40	Deep Bass
40	100	Midbass
100	250	Upper Bass
250	500	Lower Midrange
500	1000	(Middle) Midrange
1000	2000	Upper Midrange
2000	3500	Lower Treble
3500	6000	Middle Treble
6000	10,000	Upper Treble
10,000	20,000	Top Octave

This rough guide will help you understand the following terms and definitions. A full
characterization of how a product "sounds" will include aspects of each of the following sonic qualities.

Tonal Balance

The first aspect of the musical presentation to listen for is the product's overall tonal balance. How well balanced are the bass, midrange, and treble? If it sounds as though there is too much treble, we call the presentation *bright*. The impression of too little treble produces a dull or rolled-off sound. If the bass overwhelms the rest of the music, we say the presentation is heavy or weighty. If we hear too little bass, we call the presentation thin, lightweight, uptilted, or lean.

A product's tonal balance is a significant—and often overwhelming—aspect of its sonic signature.

Overall Perspective

The term *perspective* describes the apparent distance between the listener and the music. Perspective is largely a function of the recording (particularly the distance between the performers and the microphones), but is also affected by components in the playback system. Some products push the presentation forward, toward the listener; others sound more distant, or *laid-back*. The forward product presents the music in front of the loudspeakers; the laid-back product makes the music appear slightly behind the loudspeakers. Put another way, the forward product sounds as though the musicians have taken a few steps toward you; the laid-back product gives the impression that the musicians have taken a few steps back.

Another way of describing perspective is by row number in a concert hall. Some products seem to "seat" the listener at the front of the hall—in Row D, say. Others give you the impression that you're sitting farther back; say, in Row S. Several other terms describe perspective. *Dry* generally means lacking reverberation and space, but can also apply to a forward perspective. Other watchwords for a forward presentation are *immediate*, *incisive*, *vivid*, *aggressive*, and *present*. Terms associated with laid-back include *lush*, *easygoing*, and *gentle*.

Products with a forward presentation produce a greater sense of an instrument's *presence* before you, but can quickly become fatiguing. Conversely, if the presentation is too laid-back, the music is uninvolving and lacking in immediacy.

A laid-back presentation invites the listener in, pulling her gently forward *into* the music, allowing her the space to explore its subtleties. It's like the difference between having a conversation with someone who is aggressive, gets in your face, and talks too loudly, compared with someone who stands back, speaking quietly and calmly.

In loudspeakers, perspective is often the result of a peak or dip in the midrange (a peak is too much energy, a dip is too little). In fact, the midrange between 1kHz and 3kHz is called the *presence region* because it provides a sense of presence and immediacy. The harmonics of the human voice span the presence region; thus, the voice is greatly affected by a product's perspective.

Chapter 4

The Treble

Good treble is essential to high-quality music reproduction. In fact, many otherwise excellent audio products fail to satisfy musically because of poor treble performance.

The treble characteristics we want to avoid are described by the terms *bright, tizzy, forward, aggressive, hard, brittle, edgy, dry, white, bleached, wiry, metallic, sterile, analytical, screechy,* and *grainy.* Treble problems are pervasive; look how many adjectives we use to describe them.

If a product has too much apparent treble, it overstates sounds that are already rich in high frequencies. Examples are overemphasized cymbals, excessive sibilance (*s* and *sh* sounds) in vocals, and violins that sound thin. A product with too much apparent treble is called bright. Brightness is a prominence in the treble region, primarily between 3kHz and 6kHz. Brightness can be caused by a rising frequency response in loudspeakers, or by poor electronic design. Many CD players and solid-state amplifiers that measure as having a flat (accurate) frequency response nevertheless add prominence to the treble.

Tizzy describes too much upper treble (6kHz–10kHz), characterized as a *whitening* of the treble. Tizzy cymbals have an emphasis on the upper harmonics, the sizzle and air that rides over the main cymbal sound. Tizziness gives cymbals more of an *sssss* than a *sssshhhh* sound.

Forward, if applied to treble, is very similar to *bright*; both describe too much treble. A forward treble, however, also tends to be dry, lacking space and air around it.

Many of the terms listed above have virtually identical meanings. *Hard, brittle,* and *metallic* all describe an unpleasant treble characteristic that reminds one of metal being struck. In fact, the unique harmonic structure created from the impact of metal on metal is very similar to the distortion introduced by a solid-state power amplifier when it is asked to play louder than it is capable of playing.

A particularly annoying treble characteristic is *graininess*. Treble grain is a coarseness overlaying treble textures. I notice it most on solo violin, massed violins, flute, and female voice. On flute, treble grain is recognizable as a rough or fuzzy sound that seems to ride on top of the flute's dynamic envelope. (That is, the grain follows the flute's volume.) Grain makes violins sound as though they're being played with hacksaw blades rather than bows—a gross exaggeration, but one that conveys the idea of the coarse texture added by grain.

The most common sources of these problems are, in rough order of descending magnitude: tweeters in loudspeakers, overly reflective listening rooms, digital source components (usually the CD player or digital processor), preamplifiers, power amplifiers, cables, and dirty AC power sources.

So far, I've discussed only problems that emphasize treble. Some products tend to make the treble softer and less prominent than live music. This characteristic is often designed into the product, either to compensate for treble flaws in other components in the system, or to make the product sound more palatable. Deliberately softening the treble is the designer's shortcut; if he can't get the treble right, he just makes it less offensive by softening it.

The following terms, listed in order of increasing magnitude, describe good treble performance: *smooth, sweet, soft, silky, gentle, liquid,* and *lush.* When the treble becomes overly smooth, we say it is *romantic, rolled-off,* or *syrupy.* A treble described as "smooth, sweet, and silky" is being complimented; "rolled-off and syrupy" suggests that the component goes too far in treble smoothness, and is therefore *colored.*

A rolled-off and syrupy treble may be blessed relief after hearing bright, hard, and grainy treble, but it isn't musically satisfying in the long run. Such a presentation tends to become *bland, uninvolving, slow, thick, closed-in,* and *lacking detail.* All these terms describe the effects of a treble presentation that errs too far on the side of smoothness. The presentation will lack *life, air, openness, extension,* and a sense of *space* if the treble is too soft. The music sounds closed-in rather than being big and open.

The best treble presentation is one that sounds most like real music. It should have lots of energy—cymbals can, after all, sound quite aggressive in real life—yet not have a synthetic, grainy, or dry character. We don't hear these characteristics in live music; we shouldn't hear them in reproduced music. More important, the treble should sound like an integral part of the music, not a detached noise riding on top of it. If a component has a colored treble presentation, however, it is far less musically objectionable if it errs on the side of smoothness rather than brightness.

The Midrange

J. Gordon Holt, *Stereophile* magazine founder and the father of observational audio equipment evaluation, once wrote, "If the midrange isn't right, nothing else matters."

The midrange is important for several reasons. First, most of the musical energy is in the midrange, particularly the important lower harmonics of most instruments. Not only does this region contain most of the musical energy, but the human ear is much more sensitive to midrange and lower treble than to bass and upper treble. Specifically, the ear is most sensitive to sounds between about 800Hz and 3kHz, and to small changes in both volume and frequency response within this band. The ear's threshold of hearing—i.e., the softest sound we can hear—is dramatically lower in the midband than at the frequency extremes. We've developed this additional midband acuity probably because the energy of most of the sounds we heard every day for hundreds of thousands of years—the human voice, rustling leaves, the sounds of other animals—is concentrated in the midrange.

Midrange colorations can be extremely annoying. Loudspeakers with peaks and dips in the mids sound very unnatural; the midrange is absolutely the worst place for loudspeaker imperfections. Confining our discussion to loudspeakers for the moment, midrange colorations overlay the music with a common characteristic that emphasizes certain sounds. The male speaking voice is particularly revealing of midrange anomalies, which are often described by comparisons with vowel sounds. A particular coloration may impart an *aaww* sound; a coloration lower in frequency may emphasize *ooohhh* sounds; a higher-pitched coloration may sound like *eeeee;* another coloration might sound *hooty*.

Some midrange colorations can be likened to the sound of someone speaking through cupped hands. Try reading this sentence while cupping your hands around your mouth. Open and close your hands while listening to how the sound of your voice changes. That's the kind of midrange coloration we sometimes hear from loudspeakers—particularly mass-market ones.

In short, if recordings of male speaking voice sound monotonous, tiring, and resonant, it's probably the result of peaks and dips in the loudspeaker's frequency response. (These colorations are most apparent on male voice when listening to just one loudspeaker.)

Terms to describe poor midrange performance include *peaky, colored, chesty, boxy, nasal, congested, honky,* and *thick*. *Chesty* describes a lower-midrange coloration that makes vocalists sound as though they have colds. *Boxy* refers to the impression that the sound is coming out of a box instead of existing in open space. *Nasal* is usually associated with an excess of energy that spans a narrow frequency range, producing a sound similar to talking with your nose pinched. *Honky* is similar to nasal, but higher in frequency and spanning a wider frequency range.

As described previously under "Perspective," too much midrange energy can make the presentation seem forward and "in your face." A broad dip in the midrange response (too little midrange energy over a wide frequency span) can give an impression of greater distance between you and the presentation.

When choosing loudspeakers, be especially attuned to the midrange colorations described. What is a very minor—even barely noticeable—problem heard during a brief audition can turn into a major irritant over extended listening.

The preceding descriptions apply primarily to midrange problems introduced by loudspeakers. Expanding the discussion to include electronics (preamps and power amps) and source components (LP playback or a digital source) introduces different aspects of midrange performance that we should be aware of.

An important factor in midrange performance is how instrumental textures are reproduced. *Texture* is the physical impression of the instrument's sound—its fabric rather than its tone. The closest musical term for texture is timbre, defined by *Merriam Webster's Collegiate Dictionary, Tenth Edition* as "the quality given to a sound by its overtones; the quality of tone distinctive of a particular singing voice or instrument." Sonic artifacts added by electronics often affect instrumental and vocal textures.

The term *grainy*, introduced in the description of treble problems, also applies to the midrange. In fact, midrange grain can be more objectionable than treble grain. Midrange grain is characterized by a coarseness of instrumental and vocal textures; the instrument's texture is granular rather than smooth.

Midrange textures can also sound hard and brittle. Hard textures are apparent on massed voices; a choir sounds *glassy, shiny,* and *synthetic*. This problem gets worse as the choir's volume increases. At low levels, you may not hear these problems. But as the choir swells, the sound becomes hard and irritating. Piano is also very revealing of hard midrange textures, the higher notes sounding annoyingly *brittle*. When the midrange lacks these unpleasant artifacts, we say the textures are *liquid*, *smooth*, *sweet*, *velvety*, and *lush*.

The Bass

Bass performance is the most misunderstood aspect of reproduced sound, among the general public and hi-fi buffs alike. The popular belief is that the more bass, the better. This is reflected in ads for "subwoofers" that promise "earthshaking bass" and the ability to "rattle pant legs and stun small animals." The ultimate expression of this perversity is boom trucks that have absurd amounts of extraordinarily bad bass reproduction.

But we want to know how the product reproduces *music*, not earthquakes. What matters to the music lover isn't *quantity* of bass, but the *quality* of that bass. We don't just want the physical feeling that bass provides; we want to hear subtlety and nuance. We want to hear precise pitch, lack of coloration, and the sharp attack of plucked acoustic bass. We want to hear every note and nuance in fast, intricate bass playing, not a muddled roar. If Ray Brown, Stanley Clarke, John Patitucci, Dave La Rue, Dave Holland, or Eddie Gomez is working out, we want to hear *exactly* what they're doing. In fact, if the bass is poorly reproduced, we'd rather not hear much bass at all.

Correct bass reproduction is essential to satisfying musical reproduction. Low frequencies constitute music's tonal foundation and rhythmic anchor. Unfortunately, bass is difficult to reproduce, whether by source components, power amplifiers, or—especially—loudspeakers and rooms.

Perhaps the most prevalent bass problem is lack of *pitch definition* or *articulation*. These two terms describe the ability to hear bass as individual notes, each having an attack, a decay, and a specific pitch. You should hear the texture of the bass, whether it's the sonorous resonance of a bowed double bass or the unique character of a Fender Precision. Low frequencies contain a surprising amount of detail when reproduced correctly.

When the bass is reproduced without pitch definition and articulation, the low end degenerates into a dull roar underlying the music. You hear low-frequency content, but it isn't musically related to what's going on above it. You don't hear precise notes, but a blur of sound—the dynamic envelopes of individual instruments are completely lost. In music in which the bass plays an important rhythmic role—rock, electric blues, and some jazz—the bass guitar and kick drum seem to lag behind the rest of the music, putting a drag on the rhythm. Moreover, the kick drum's dynamic envelope (what gives it the sense of sudden impact) is buried in the bass guitar's sound, obscuring its musical contribution. These conditions are made worse by the common mid-fi affliction of too much bass.

Terms descriptive of this kind of bass include *muddy*, *thick*, *boomy*, *bloated*, *tubby*, *soft*, *congested*, *loose*, and *slow*.

Chapter 4

Terms that describe excellent bass reproduction include *taut, quick, clean, articulate, agile, tight,* and *precise.* Good bass has been likened to a trampoline stretched taut; poor bass is a trampoline hanging slackly.

The amount of bass in the musical presentation is very important; if you hear too much, the music is overwhelmed. Excessive bass is a constant reminder that you're listening to reproduced music. This overabundance of bass is described as *heavy.*

If you hear too little bass, the presentation is *thin, lean, threadbare,* or *overdamped.* An overly lean presentation robs music of its rhythm and drive—the full, purring sound of bass guitar is missing, the depth and majesty of double bass or cello are gone, and the orchestra loses its sense of power. Thin bass makes a double bass sound like a cello, a cello like a viola. The rhythmically satisfying weight and impact of bass drum are reduced to shadows of their former power. Instruments' harmonics are emphasized in relation to the fundamentals, giving the impression of well-worn cloth that's lost its supporting structure. A thin or lean presentation lacks *warmth* and *body.* As described earlier in this chapter in the discussion of audio sins of commission and omission, an overly lean bass is preferable to boomy bass.

Two terms related to what I've just described about the quantity of bass are *extension* or *depth.* Extension is how deep the bass goes—not the bass and upper bass described by *lean* or *weighty,* but the very bottom end of the audible spectrum. This is the realm of kick drum and pipe organ. All but the very best systems *roll off* (reduce in volume) these lowermost frequencies. Fortunately, deep extension isn't a prerequisite to high-quality music reproduction. If the system has good bass down to about 35Hz, you don't feel that much is missing. Pipe-organ enthusiasts, however, will want deeper extension and are willing to pay for it. Reproducing the bottom octave correctly can be very expensive.

Much of music's dynamic power—the ability to convey wide differences between loud and soft—is contained in the bass. Though I'll discuss dynamics later in this section, bass dynamics bear special discussion—they are that important to satisfying music reproduction.

A system or component that has excellent bass dynamics will provide a sense of sudden impact and explosive power. Bass drum will jump out of the presentation with startling power. The dynamic envelope of acoustic or electric bass is accurately conveyed, allowing the music full rhythmic expression. We call these components *punchy,* and use the terms *impact* and *slam* to describe good bass dynamics.

A related aspect is *speed,* though, as applied to bass, "speed" is somewhat of a misnomer. Low frequencies inherently have slower attacks than higher frequencies, making the term technically incorrect. But the *musical* difference between "slow" and "fast" bass is profound. A product with fast, tight, punchy bass produces a much greater rhythmic involvement with the music. (This is examined in more detail later.)

Although reproducing the sudden attack of a bass drum is vital, equally important is a system's ability to reproduce a fast decay; i.e., how a note ends. The bass note shouldn't continue after a drum whack has stopped. Many loudspeakers store energy in their mechanical structures and radiate that energy slightly after the note itself. When this happens, the bass has *overhang*, a condition that makes kick drum, for example, sound *bloated* and *slow*. Music in which the drummer used double bass drums is particularly revealing of bass overhang. If the two drums merge into a single sound, overhang is probably to blame. You should hear the attack and decay of each drum as distinct entities. Components that don't adequately convey the sudden dynamic impact of low-frequency instruments rob music of its power and rhythmic drive.

Soundstaging

Soundstaging is the apparent physical size of the musical presentation. When you close your eyes in front of a good playback system, you can "see" the instrumentalists and singers before you, often existing within an acoustic space such as a concert hall. The soundstage has the physical properties of width and depth, producing a sense of great size and space in the listening room. Soundstaging overlaps with imaging, or the way instruments appear as objects hanging in three-dimensional space within the recorded acoustic. As mentioned previously in this chapter, a large and well-defined soundstage is most often heard when playing audiophile-grade recordings made in a real acoustic space such as a concert hall or church.

The most obvious descriptions of the soundstage are its physical dimensions—width and depth. You hear the musical presentation as existing beyond the left and right loudspeaker boundaries, and extending farther away from you than the wall behind the loudspeakers.

Of all the ways music reproduction is astounding, soundstaging is without question the most miraculous. Think about it: The two loudspeakers are driven by two-dimensional electrical signals that are nothing more than voltages that vary over time. From those two voltages, a huge, three-dimensional panorama unfolds before you. You don't hear the music as a flat canvas with individual instruments fused together; you hear the first violinist to the left front of the presentation, the oboe farther back and toward the center, the brass behind the basses on the right, and the tambourine behind all the other instruments at the very rear. The sound is made up of *individual objects* existing within a space, just as you would hear at a live performance. Moreover, you hear the oboe's timbre coming from the oboe's position, the violin's timbre coming from the violin's position, and the hall reverberation surrounding the instruments. The listening room vanishes, replaced by the vast space of the concert hall—all from two voltages.

A soundstage is created in the brain by the time and amplitude differences encoded in the two audio channels. When you hear instrumental images toward the rear right of the soundstage, the ear/brain is synthesizing those aural images by processing the slightly different information in the two signals arriving at your ears. Visual perception works the same way: there is no depth information present on your retinas; your brain extrapolates the appearance of depth from the differences between the two flat images.

Audio components vary greatly in their abilities to present these spatial aspects of music. Some products shrink soundstage width and shorten the impression of depth. Others reveal the glory of a fully developed soundstage. I find good sound-stage performance crucial to satisfying musical reproduction. Unfortunately, many products destroy or degrade the subtle cues that provide soundstaging.

Terms descriptive of poor soundstage width are *narrow* and *constricted*—the music, squeezed together between the loudspeakers, does not envelop the listener. A soundstage lacking depth is called *flat*, *shallow*, or *foreshortened*. Ideally, the soundstage should maintain its width over its entire depth. A soundstage that narrows toward the presentation's rear robs the music of its size and space.

The illusion of soundstage depth is aided by resolution of low-level spatial cues such as hall reflections and reverberation. In particular, the reverberation decay after a loud climax followed by a rest helps define the acoustic space. The loud signal is like a flash of light in a dark room; the space is momentarily illuminated, allowing you to see its dimensions and characteristics.

Now that we've covered space and depth, let's discuss how the instrumental images appear within this space. Images should occupy a specific spatial position in the soundstage. The sound of the bassoon, for example, should appear to emanate from a specific point in space, not as a diffuse and borderless image. The same could be said for guitar, piano, sax, or any other instrument in any kind of music. The lead vocal should appear as a tight, compact, definable point in space exactly between the loud-speakers. Some products, particularly large loudspeakers, distort image size by making every instrument seem larger than life—a classical guitar suddenly sounds ten feet wide. A playback system should reveal somewhat correct image size, from a 60'-wide symphony orchestra to a solo violin. I say "somewhat" because it is impossible to re-create the correct spatial perspectives of such widely divergent sound sources through two loudspeakers spaced about 8' apart. Although image size and placement are char-acteristics inherent in the recording, they are dramatically affected by components in the playback system.

Terms that describe a clearly defined soundstage are *focused*, *tight*, *delineated*, and *sharp*. *Image specificity* also describes tight image focus and pinpoint spatial accuracy. A poorly defined soundstage is described as *homogenized*, *blurred*, *confused*, *congested*, *thick*, and lacking *focus*.

Some products produce a crystal-clear, *see-through* soundstage that allows the listener to hear all the way to the back of the hall. Such a *transparent* soundstage has a lifelike immediacy that makes every detail clearly audible. Conversely, an *opaque* sound-stage is *thick* or *murky*, with less of an illusion of "seeing" into space. *Veiling* is often used to describe a lack of transparency.

Finally, superb soundstaging is relatively fragile. You need to sit directly between the loudspeakers, and every component in the playback chain must be of high quality. Soundstaging is easily destroyed by low-quality components, a bad listening room, or poor loudspeaker placement. This isn't to say you have to spend a fortune to get good soundstaging; many very-low-cost products do it well, but it is more of a challenge to find those bargains.

Dynamics

The dynamic range of an audio system isn't how loudly it will play, but rather the difference in level between the softest and loudest sounds that the system can reproduce. It is often specified technically as the difference between the component's noise level and its maximum output level. A symphony orchestra has a dynamic range of about 100 decibels, or dB; a typical rock recording's dynamic range is about 10dB. In other words, comparatively speaking, the rock band is always loud; it has little dynamic range.

Dynamics are a very important part of music reproduction. They propel the music forward and involve the listener. Much of music's expression is conveyed by dynamic contrast, from *pp* (pianissimo) to *fff* (triple forte).

There are two distinct kinds of dynamics. *Macrodynamics* refers to the presentation's overall sense of slam, impact, and power—bass-drum whacks and orchestral crescendos, for examples. If the system has poor macrodynamics, we say the sound is *compressed* or *squashed*.

Microdynamics occur on a smaller scale. They don't produce a sense of impact, but are essential to providing realistic dynamic reproduction. Microdynamics describe the fine dynamic structure in music, from the attack of a triangle or other small percussion instruments in the back of the soundstage to the suddenness of a plucked string on an acoustic guitar. Neither sound is very loud in level, but both have dynamic structures that require agility and speed from the playback system.

Products with good dynamics—macro and micro—make the music come alive, allowing a vibrancy and life to emerge. Dynamic changes are an important vehicle of musical expression; the more you hear the musicians' intent, the greater the musical communication between performers and listener. Some otherwise excellent components fail to convey the broad range of dynamic contrast.

These characteristics are associated with *transient response*, a system's ability to quickly respond to an input signal. A transient is a short-lived sound, such as that made by percussion instruments. Transient response describes an audio system's ability to faithfully reproduce the quickness of transient signals. For example, a drum being struck produces a waveform with a very steep attack (the way the sound begins) and a fast decay (the way a sound stops). If any component in the playback system can't respond as quickly as the waveform changes, a distortion of the music's dynamic envelope occurs, and the steepness is slowed. Audio components described as *quick* or *fast* reproduce the suddenness of transient signals.

But just because a component or system can reproduce loud and soft levels doesn't necessarily mean it has good dynamics. We're looking for more than a wide dynamic range. The system must be capable of expressing fine gradations of dynamics, not just loud and soft. As the music changes in level (which, except in many rock recordings, it's doing most of the time), you should hear loudness changes along a smooth continuum, not as abrupt jumps in levels.

Chapter 4

Detail

Detail refers to the small or low-level components of the musical presentation. The fine inner structure of an instrument's timbre is one kind of detail. The term is also associated with transient sounds (those with a sudden attack) at any level, such as those made by percussion instruments. A playback system with good resolution of detail will infuse music with that sense that there is simply more music happening.

Assembling a good-sounding music system or choosing between two components can often be a tradeoff between smoothness and the resolution of detail. Many audio components hype detail, giving transient signals an *etched* character. Etch is an unpleasant hardness of timbre on transients that emphasizes their prominence. Sure, you can hear all the information, but the presentation becomes too *aggressive, analytical,* and *fatiguing:* low-level information is brought up and thrust at you, and you feel a sense of relief when the music is turned down or off—not a good sign.

Components that err in the opposite direction don't have this etched and analytical quality, but neither do they resolve all the musical information in the recording. These components are described as overly *smooth,* or having *low resolution.* They tend to make music bland by removing parts of the signal needed for realistic reproduction. These kinds of components don't rivet your attention on the music; they are *uninvolving* and dull. You aren't offended by the presentation, as you are with an analytical system, but something is missing that you need for musical satisfaction.

It is a rare product indeed that presents a full measure of musical detail without sounding etched. The best products will reveal all the low-level cues that make music interesting and riveting, but not in a way that results in listening fatigue—that sense of tiredness after a long listening session. The music playback system must walk the very fine line between resolution of real musical information and sounding analytical.

Musicality

Finally, we get to the most important aspect of a system's presentation—musicality. Unlike the previous characteristics, musicality isn't any specific quality that you can listen for, but the overall musical satisfaction the system provides. Your sensitivity to musicality is destroyed when you focus on a certain aspect of the presentation; i.e., when you listen critically. Instead, musicality is the gestalt, the whole of your reaction to the reproduced sound. We also use the term *involvement* to describe this oneness with the music. A sure indication that a component or a system has musicality is when you sit down for an analytical listening session and minutes later find yourself immersed in the music and abandoning the critical listening session. This has happened many times to me as a reviewer, and is a good measure of the product's fundamental "rightness." Ultimately, musicality—not dissecting the sound—is what high-end audio is all about.

5

Digital Source Components

Not that long ago, a discussion of digital source components was confined to a single format: the compact disc. But today's music lover is faced with a proliferation of new digital formats, from terrific-sounding high-resolution media such as SACD and DVD-Audio, to lower-quality carriers such as MP3 players and computer-based audio. Technology has also changed the way we access digital music; hard-disk-based music servers now give us instant access to our music libraries without removing a disc from its case and inserting the disc in a player. The iPod and other portable music players allow us to take our music collections with us wherever we go.

This chapter will guide you through the various formats for accessing digital music, explain how they work, and provide advice on how to choose the best digital components for your system and budget.

Compact Disc

When the compact disc was introduced in 1982, many music lovers found its sound inferior to the "crude" technology of vinyl records. CD's marketing slogan, "Perfect Sound Forever," seemed like a cruel joke. But despite the fundamental limitations of CD, its sound quality has steadily improved to the point where it now can be musical and enjoyable. In fact, just since the turn of the millennium, CDs have taken a big leap forward, the result of higher-quality mastering formats and techniques. Despite the explosion of new digital-audio formats since the CD's introduction in 1982, the CD continues to be the primary medium by which most of us listen to music.

But getting the best sound from your CD collection requires a player capable of extracting all the information on the disc and presenting it in a musical way. The differences in sound quality between players is significant, and choosing the best player for your system is an important aspect of building a system that will give you years of musical enjoyment.

Although the sound quality on compact discs themselves has greatly improved, the quality of CD playback hardware has, in one respect, declined over the past few years. This trend is the result of the explosion of the DVD format and the staggering price-to-performance ratio of today's DVD players. Video-processing circuitry became simultaneously more advanced and vastly less expensive, due to large-scale manufacturing of integrated circuits. In fact, the same video-processing circuitry that once cost $10,000 in a stand-alone video processor is routinely available in DVD players costing well under $100.

But how does this trend toward better and cheaper DVD players affect the sound quality of CD playback?

Chapter 5

Most consumers today buy a single machine that plays DVDs, CDs, and perhaps other disc formats. For many, the DVD player serves as the source for CD listening. As DVD players got cheaper and cheaper in a highly competitive market segment, the designers cut corners on the audio circuitry. The parts in a DVD player that affect sound quality when playing CDs include the power supply (a large and heavy section), the audio amplifiers, passive components (primarily resistors and capacitors), and a host of other devices. These parts have become *more* expensive over time, not less so. Consequently, a $100 DVD player's audio-parts-quality is abysmal, and the sound quality often reflects this fact.

The solution is to choose either a dedicated high-end CD player, or a DVD or multi-format *universal* player with high-end parts and design. Many specialty audio companies offer DVD players with outstanding audio sections that were not thrown in as afterthoughts, or put together with whatever parts budget was left over after the video circuitry was designed. Despite this global trend toward lower-quality CD playback, high-end CD players have never sounded better, or been a better value than they are today.

High-End CD Playback: Dedicated CD Players and Separate Transports and Processors

As previously discussed, a dedicated CD player (one that doesn't play other formats) will generally deliver better sound quality than is possible from a DVD player or universal machine. There are some outstanding universal players, but they are the exception rather than the rule.

If your focus is on 2-channel music listening, you'll probably choose a CD-only playback system. If so, you have two choices: a CD player or separate CD transport and digital-to-analog converter (also called a digital processor). A CD transport is simply a CD player without digital-to-analog conversion circuitry. It outputs a digital signal that is converted to analog by the outboard digital processor, housed in a separate chassis. Needless to say, this approach is considerably more expensive than buying a CD player.

But are "separates" worth the price? If you have a large budget and want the best possible performance, the answer is yes. But CD players are generally a better value for two reasons: 1) By combining the transport and processor in one chassis, with one power supply, one front panel, one shipping carton, and one AC cord, the manufacturer can put more of the manufacturing budget into better sound; and 2) A CD player has no need of a sonically degrading digital interface (cable) between its transport and processor, which means that it might have better sound.

A CD player also makes your life much simpler than owning separates. Rather than requiring two chassis in your rack, two power cords, and a digital interconnect, the CD player lets you focus on the music rather than on the equipment.

Many excellent CD players are available for under $500, including some legitimate high-end machines that sell for as little as $300. This is, however, the lower limit of true high-end CD players. Below this level you enter the realm of mass-market products designed for maximum features and minimum manufacturing cost, not musical performance. You should also avoid multi-disc players; most manufacturers cut corners in sound quality to cover the extra expense of the disc-changing mechanism. The single-disc player will provide better sound quality for the same price, and also better ultimate fidelity—single-disc transport mechanisms can be made better than carousel-type changers. If, however, you're set on a changer, consider using it as a transport to feed a high-quality outboard digital-to-analog converter. A few CD changers can be called high-end, but they are rare.

Separates offer several advantages over one-box CD players. First, they allow you to upgrade the transport and the processor independently. Second, putting the sensitive digital-to-analog conversion electronics in a separate chassis from the mechanical transport lowers the chance of polluting the analog signal with digital noise. Third, the cutting-edge design work takes place in separates, not CD players.

Those arguments, however, are less compelling in light of what's been happening in CD-player design recently. Designers are now applying the same level of parts and design quality to CD players once reserved for separates. These premium parts and techniques include large and well-regulated power supplies, high-quality DAC (digital-to-analog converter) chips, and the designer's best analog circuits. Moreover, many innovative design techniques are starting to appear first in CD players. The CD player is no longer automatically the entry-level product in the digital manufacturer's line, instead it's become worthy of the manufacturer's best efforts.

High-Definition Compatible Digital (HDCD)

Some CD and universal (multi-format) disc players are touted as being "HDCD Compatible." This designation means the player will decode CDs encoded with the High-Definition Compatible Digital process. Recordings encoded with HDCD can be decoded and played back on any CD player, with some improvement in fidelity. But if your CD player is equipped with an HDCD decoder, HDCD recordings can be reproduced with much better sound quality than is possible with conventional CDs.

Digital Connections

The S/PDIF (Sony/Philips Digital Interface Format) interface is a method of transmitting digital audio from one component to another. For example, a digital transport's S/PDIF output is carried down a digital interconnect to the S/PDIF input on a digital processor.

Digital connections can be divided into two categories: electrical and optical. In an electrical connection, electrons carry the signal down copper or silver wire. An optical connection transmits light down a plastic or glass tube. Both types carry the same S/PDIF format, but in different ways.

The most common type of connection, *coaxial*, is carried on an RCA cable. This is the electrical connection found on virtually all CD transports, most good CD players, and other consumer digital audio products such as digital preamplifiers, room-correction systems, and digital recorders. *TosLink* is the low-cost optical interface promoted by mass-market audio manufacturers as an alternative to coaxial connection. TosLink, a trademarked name of the Toshiba Corporation, is more properly called "EIAJ Optical," after the Electronics Industries Association of Japan. Fig.5-1 shows coaxial and TosLink digital outputs as you would find them on a CD transport, DVD player, or universal disc machine.

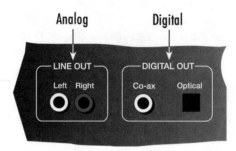

Fig.5-1 A digital source's output can be on an electrical coaxial output (left) or on an optical TosLink output (right). (Courtesy Dolby Laboratories)

Note that coaxial and TosLink interconnects can carry S/PDIF from a CD transport to a digital-to-analog converter (or from a CD player to an A/V receiver), as well as a Dolby Digital or DTS bitstream from a DVD player, satellite receiver, cable box, or other audio/video source to an A/V receiver. If you connect a coaxial or TosLink cable carrying Dolby Digital to a digital-to-analog converter that expects to see S/PDIF from a CD transport, you'll hear a blast of noise.

There's a third interface, found only on upper-end components, called the *AES/EBU* interface, which is carried on a balanced line terminated with three-pin XLR connectors. Of the three conductors in a balanced signal, one is ground, one is the digital signal, and the third is the digital signal inverted. Virtually all transports and processors have RCA jacks, and most include TosLink for compatibility. AES/EBU is usually found on only the more expensive components, or offered as an option on mid-priced digital equipment. Fig.5-2 shows these interface types.

TosLink is the least good interface, mechanically (the physical connection between cable and jack), electrically, and sonically. TosLink connection tends to blur the separation between individual instrumental images, adds a layer of grunge over instrumental textures, softens the bass, and doesn't have the same sense of black silence between notes. My advice is to avoid TosLink unless you have components equipped only with TosLink connections, or you've run out of coaxial digital inputs on your AVR.

But how can a digital interface, that merely transmits ones and zeros, change the sound? If the ones and zeros are the same (which they are), doesn't the sound have to be the same? How can a digital system, where the music is represented by ones and zeros, exhibit an analog-like variability in sound quality?

Fig.5-2 Digital signals are carried on cables terminated with RCA connectors (left), TosLink connectors (middle), or XLR connectors (right). (Courtesy Monster Cable Products)

The answer is that the digital interface makes no sonic difference if the audio data are being transported from one device to another and not converted to analog for listening. The problem arises when we *listen* to digital audio that has been put through a digital interface because the interface introduces *jitter*. The term jitter refers to timing errors in the clock that controls when the CD's digital samples are converted into music. Musicians say that "The right note at the wrong time is the wrong note," and this perfectly describes the jitter problem in digital-audio reproduction. If the digital samples are converted into an analog signal (music) with a jittered clock, a form of degradation is introduced.

High-Resolution Digital Audio: SACD and DVD-Audio

Despite the advances made in the past few years, CD sound quality is fundamentally compromised by its sampling rate of 44.1kHz and its word length of 16 bits. The sampling rate determines the highest audio frequency that can be encoded, and the word length determines the system's dynamic range, noise level, and resolution. Digital audio systems with higher sampling rates and longer word length sound significantly better than CD, whose parameters reflect late-1970's technology.

High-resolution digital audio is generally defined as any system with a sampling rate of 88.2kHz or greater, and word length of 18 bits or greater. The highest sampling rate and word length in use today is 192kHz/24-bit. That may seem overkill, but increasing the sampling rate and word length greatly improves the sound of digitally reproduced music. High-resolution digital audio has greater resolution of low-level detail, wider dynamic contrasts, more natural rendering of timbre, smoother treble, and a larger and more spacious soundstage.

Audiophiles have two choices in high-resolution digital: Super Audio CD (SACD) and DVD-Audio. Neither format became the mass-market successor to CD, but both still exist as niche formats. As of this writing, there are about 4000 SACD titles and 800 DVD-A titles available. If you choose to buy an SACD or DVD-A player, be aware that the formats will never generate a huge catalog of titles. New titles will be released slowly, and be primarily of classical music. Nonetheless, the sound quality from SACD and DVD-A is so good that it's worth buying a player, in my view. Moreover, if you want high-quality multichannel audio in your home, there's no substitute. There are enough wonderful high-res titles on the market to warrant investing in a quality machine. For a complete list of SACD titles, go to www.sa-cd.net.

SACD

Let's first look at SACD. The format provides up to 74 minutes of 2-channel and 6-channel audio (6-channel is optional) on a disc the size of a CD. Where CD has a bandwidth of 20kHz, SACD's bandwidth is 100kHz. SACD also has wider dynamic range than CD, with greater resolution of low-level detail. SACD doesn't use conventional *pulse-code modulation* (PCM) employed on CD and DVD-Audio, but a different encoding technique called Direct Stream Digital (DSD).

SACD also offers the possibility (but not the requirement) of a "hybrid" disc containing two layers on the same disc for backward-compatibility with CD. One layer contains CD's conventional 16-bit/44.1kHz audio and the second layer holds the high-resolution multichannel version. When you put a hybrid SACD disc in a CD player, it behaves (and sounds) just like a CD. But put the same disc in an SACD machine and you can hear high-resolution stereo or multichannel music. The hybrid disc ensures backward-compatibility with the massive installed base of CD players. This feature is made possible by two information-carrying layers on the disc, one high-resolution and one that is CD-compatible (Fig.5-3). The CD-quality layer is sometimes called the *Red Book* layer, after the color of the cover of the official specification describing the compact disc. More information is contained in the second layer because the information-carrying pits are about half the size of a CD's pits, are spaced closer together, and the information is recovered with a shorter wavelength playback laser (Fig.5-4).

Note that some SACD players are 2-channel only; others will play both 2-channel and multichannel recordings. SACD playback is available in combination CD/SACD machines, as well as in some universal-disc players.

DVD-Audio

The DVD-Video format has been a massive commercial success, rapidly replacing VHS as the preferred movie-delivery format. As mentioned earlier in this chapter, DVD's larger data capacity (nearly 18 gigabytes in its maximum form, 25 times that of

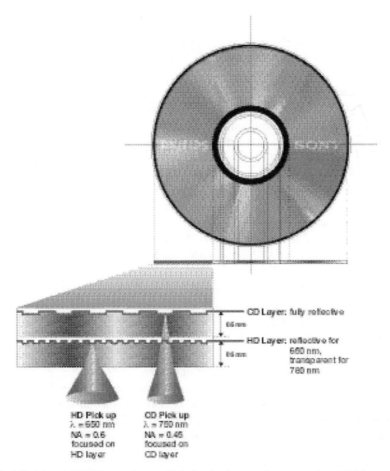

CD Layer: fully reflective

HD Layer: reflective for
650 nm,
transparent for
780 nm

HD Pick up
λ = 650 nm
NA = 0.6
focused on
HD layer

CD Pick up
λ = 780 nm
NA = 0.45
focused on
CD layer

Fig.5-3 A hybrid SACD contains a a high-density layer and a conventional CD layer on the same disc. (Courtesy Sony)

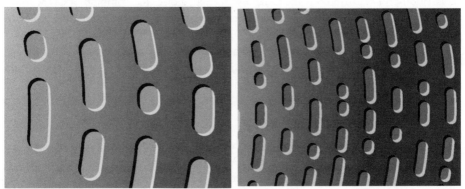

Fig.5-4 An SACD's information-carrying pits (right) are smaller and closer together than on a CD. (Courtesy Sony)

CD) makes it possible to store high-resolution audio—the DVD-Audio format. Specifically, the DVD-A specification calls for a disc that can contain 6-channel music along with a 2-channel mix on the same disc, with a sampling rate as high as 192kHz and word lengths of up to 24 bits. You can think of a DVD disc as a "bit bucket" that can hold a wide variety of sampling frequencies, word lengths, playing times, and number of channels, all selectable by the disc's producer.

Even DVD-A's storage capacity and maximum transfer rate (how quickly the bits can be pulled off the disc) are not high enough to deliver multichannel audio with high resolution on all channels. To overcome this limitation, DVD-A employs a lossless coding system developed by Meridian Audio called Meridian Lossless Packing (MLP). Unlike "lossy" compression systems such as Dolby Digital or MP3 that remove information and reduce fidelity, MLP is a perfectly lossless process, producing identical bit-for-bit data on playback with no sonic degradation. Using MLP is completely transparent to the user; you simply insert the DVD-A disc into player and the decoding is automatically engaged. More than 90% of all DVD-A releases are encoded with MLP.

MLP is also the basis for Dolby TrueHD, a lossless system for delivering high-resolution multichannel audio on HD DVD and Blu-ray Disc.

Using a DVD-Audio player is somewhat different from playing an SACD. It is assumed that the DVD-A player is connected to a video monitor; set-up menus, selection of 2-channel or multichannel, and other parameters are chosen through a video-based menu system. Operating a DVD-A player without a video display can be challenging. Conversely, SACD machines have no such requirement. They operate just as a CD player would; pop in a disc, press play, and start listening.

Neither SACD nor DVD-Audio became a mass-market success, probably because the two formats competed against each other, and were incompatible. We will never see a huge library of titles in either format, but many audiophiles nonetheless choose an SACD or DVD-A player to enjoy the titles that are available.

DualDisc

A variation on DVD-Audio is the DualDisc format, a disc that combines a CD with DVD. As its name suggests, a DualDisc is composed of two discs bonded together. One disc contains conventional CD data, and plays on any CD machine. The opposite side is a DVD, which can contain video, high-resolution digital audio (with the same specs as DVD-A), graphics, or data. DualDisc can offer full DVD-A-quality audio and be backward-compatible with the world's installed base of CD players.

Universal Disc Players

Many manufacturers have introduced universal players that are compatible with a wide range of disc formats, including CD, SACD, DVD-A, DVD-Video, and even discs on which MP3 files are stored. Universal players are a good solution if you want to enjoy the broadest possible range of formats and title selection.

You'll find a number of compatibility variations in universal disc players. They include:

• CD, 2-channel SACD, and 2-channel DVD-A

• CD, 2-channel and multichannel SACD, DVD-Video

• CD, 2-channel and multichannel SACD, 2-channel and multichannel DVD-A, DVD-V

• CD, 2-channel and multichannel DVD-A, DVD-Video

For those seeking the ultimate in sound quality, choosing separate machines to play each format is a better choice than the universal player. Although many offer excellent performance, they rarely equal the sound quality of dedicated single-format machines.

Multichannel Output on High-Resolution Players

All multichannel audio players (DVD-Audio, SACD, and universal machines) deliver their outputs in analog form on six jacks. This requires that you have an analog preamplifier, controller, or receiver with six analog inputs, and that you run six analog cables between the player and your preamp or controller. A few multichannel players have a digital output that carries high-resolution digital multichannel audio, but these multichannel digital interfaces are often proprietary, meaning that you'll need a controller or receiver made by the same company to use this connection. Examples of proprietary multichannel high-resolution digital connections are Sony's i.LINK and Denon's DenonLink.

Fig. 5-5 is the back panel of a multichannel SACD player showing the 2-channel and multichannel outputs. You can also see the coaxial and TosLink digital output jacks which carry 44.1kHz/16-bit digital audio to an outboard D/A converter or to a controller or A/V receiver (described earlier in this chapter). The jack marked i.LINK is the proprietary high-resolution multichannel output that requires an i.LINK-compatible receiver.

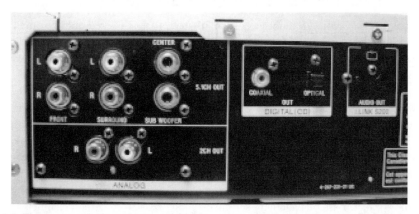

Fig.5-5 The rear panel of a multichannel SACD machine showing the analog multichannel outputs (top left), stereo analog outputs (bottom left), coaxial and TosLink outputs for CD-data only (center), and the i.LINK high-resolution multichannel output (right).

Note that most controllers and receivers offer just one set of multichannel analog-input jacks; if you have two multichannel sources, you can connect only one at a time.

The newest generation of digital interfaces, High-Definition Multimedia Interface (HDMI), carries high-resolution multichannel digital audio. The interface was developed primarily to carry high-definition video as well as audio in a single, easy-to-use cable. (An HDMI cable is shown in Fig.5-6.) The industry-wide acceptance of HDMI v1.3 solves three problems: 1) the need to run six analog cables between a multichannel source and your receiver or A/V controller; 2) the limitation of just one multichannel input on receivers and controllers; 3) the incompatibility between brands of equipment employing proprietary digital interfaces.

Fig.5-6 An HDMI cable (v1.3) can carry high-resolution multichannel digital audio from a digital source to an outboard digital-to-analog converter, A/V controller, or AVR. HDMI can also simultaneously carry high-definition video. (Courtesy Monster Cable Products)

Bass Management in SACD and DVD-A Players

The term *bass management* describes a system in which bass is filtered out of some channels and directed to a subwoofer, if the system employs one. Bass management is necessary in multichannel audio systems, and is found in most SACD and DVD-A players, as well as in A/V receivers and A/V controllers.

To see why we need bass management, consider a recording containing full-bandwidth information (low bass as well as midrange and treble) in every channel. A good example is a recording of popular music in which the bass guitar and kick drum are at least partially positioned in the center channel. Most multichannel loudspeaker arrays employ large floorstanding loudspeakers in the left and right positions, and smaller speakers in the center and rear positions. Low frequencies from the bass guitar and kick drum would overload the small center speaker, introducing distortion or damaging the speaker. Bass management selectively filters bass from certain channels you specify according to the kind of loudspeakers in your system. That low bass can be directed to a subwoofer, if your system includes one.

If you've set up a home-theater receiver or controller and selected LARGE or SMALL from the menu for each of your speakers, you've worked with bass management. The SMALL setting simply engages a high-pass (low-cut) filter on that channel, preventing low bass from reaching the speaker. A typical cutoff frequency is 80Hz.

Bass management in the DVD-A format is performed via on-screen menus. SACD players typically have a selection of common loudspeaker configurations from which to choose; you simply engage one from the player's front panel with no need for a video display. For example, setting #1 may be ideal for a system with large left and right speakers, small center and surround speakers, and a subwoofer.

The bass-management set-up controls are sometimes accompanied by a variable delay to each channel. You set the amount of delay according to how far you sit from each loudspeaker so that the sound from each speaker reaches you simultaneously. For example, if you sit 12' from the left and right speakers, but only 6' from the rear speakers, six milliseconds (6ms) of delay to the rear speakers result in coincident arrival of front sounds with those from the rear. (Sound travels at roughly one foot per millisecond.)

The Future of High-Resolution Digital Audio

Unfortunately, we may never see a consumer format that delivers high-res audio *and* a huge library of titles. The reason is that record companies believe consumers are satisfied with CD sound, and even with the inferior sound quality of MP3 and other highly compressed formats. The mass acceptance of MP3 bears this out, although many of us care about music and the quality of its reproduction.

Nonetheless, the two formats that are now vying to replace DVD as the high-definition video replacement for the DVD offer the provision for audio-only configurations that can deliver very high-quality stereo or multichannel digital audio.

The first of these formats, HD DVD, has an audio-only form in which it can store up to eight channels of digital audio with sampling rates up to 96kHz and word lengths of up to 24-bits. In 2-channel mode, the sampling rate can be increased up to 192kHz. The rival Blu-ray Disc also can serve as an audio-only carrier, with up to eight channels encoded at 192kHz sampling frequency and 24-bit resolution. Note that Blu-ray has somewhat higher specs due to its greater storage capacity (25GB per layer for Blu-ray vs. 15GB per layer for HD DVD). Whether either or both of these formats becomes a high-resolution music carrier remains to be seen.

What is not in doubt is that we will see concert videos and musical performances on these formats accompanied by uncompressed, high-resolution audio. As mentioned earlier, the Dolby TrueHD format delivers high-resolution multichannel audio to accompany the video with perfect bit-for-bit accuracy to the source. The competing DTS system also offers a lossless high-res delivery format called DTS-HD Master Audio. Both Dolby TrueHD and DTS-HD Master Audio are options to the disc producers, not requirements of the HD DVD and Blu-ray Disc formats. (Chapter 10 has a more in-depth discussion of these newer surround-sound formats and their application to home theater.)

Chapter 5

Music Servers

An entirely new product category called the *music server* made its introduction in 2002. This digital-audio recording device employs large hard-disk drives to store hundreds of hours of music, usually recorded from CD (Fig.5-7). Transferring your CD library to a single computer-based device confers many advantages, including instant access to any piece of music without searching through CDs; no need to open a CD and insert it into a player; no CD-storage problems or clutter; and the ability to create custom music playlists by genre or group. You could, for example, tell the server to play two hours of a particular artist, or six hours of classic jazz, or three hours of chamber music. The digital-audio server opens up new possibilities for the way in which we access our music (Fig.5-8). Incidentally, Apple's iPod is essentially a music server that happens to be portable and also contains a miniature stereo system.

Fig.5-7 A music server can store a large music library on its integral hard-disk drives, and provides instant access to any selection. (Courtesy ReQuest)

Fig.5-8 Music servers are operated through an on-screen graphic user interface. (Courtesy Yamaha)

Audio can be stored on the server's hard drives uncompressed (consuming about 10.5MB per stereo minute), or with a compression scheme that increases a given drive's storage capacity at the expense of sound quality. Compression systems are available that result in about a 2:1 reduction in the storage requirement, but don't degrade sound quality—they achieve perfect bit-for-bit accuracy to the source.

Some servers have fixed hard-drive capacity, with no provision for expansion if you need more storage space. Others allow you to add additional drives as your music collection grows. A few also provide the option of removable drives so that you can back-up your music. This requires copying the data from one drive to another, removing one of the drives, and putting it away in case of a drive crash. Others simply store the audio data on two separate drives. As anyone who's used a personal computer knows, the hard-disk drive is the least reliable component, and prone to sudden and catastrophic failure. If this happens to the drives in your server, you'll need to re-record (and re-classify) all of your CDs, unless you have a server with the removable backup option. If you choose to commit your music library to a digital-audio server, I strongly recommend keeping the original CDs in case of a drive crash.

Computer-Based Digital Audio and File Formats

The personal computer has revolutionized how a new generation of music listeners accesses and enjoys music. The ability to download virtually any piece of music with a few keystrokes and mouse-clicks has been both a blessing and a curse. Although easy access to a wide variety of music is generally a good thing, downloaded music is severely compromised in sound quality. The popular MP3 format, for example, encodes music with perhaps one-tenth the data rate of CD, and one-fiftieth the data rate of a high-resolution format such as DVD-Audio. MP3 and other such "lossy" compression formats were designed for convenience and low transmission and storage requirements, not sound quality. Sadly, many young people today have never heard their favorite music via an uncompressed format.

Not all computer-based audio is low-resolution; the Music Giants download service offers high-resolution downloads. Many portable audio players that rely on a computer, such as Apple's iPod, have the ability to store uncompressed music, or music using a lossless compression system.

The iPod can be part of a high-end audio system provided that you also have a CD player for loading music onto it (as opposed to downloading music) and that you store the music using either WAV files or Apple Lossless encoding. WAV files are identical to the source data on the CD; there is no loss in fidelity. The downside is that WAV files consume lots of disk-storage space (about 10.5MB per stereo minute). A better alternative is Apple Lossless, an encoding scheme that cuts in half the storage requirements with no loss in sound quality. The digital bits that come out of Apple Lossless are identical to those that went in.

Using an iPod or other portable music player as a source in a high-end system gives you instant access to your favorite music, and in effect functions as a music server. The downside is that the sound quality is compromised by the portable player's digital-to-analog conversion circuitry, as well as the quality of its analog output electronics. The iPod is, however, surprisingly good sounding, and is a great choice for budget systems or for programming background music.

How to Choose a Digital Source

When choosing a digital source, you must first decide what formats you will be listening to. If CDs will be your primary digital source, choose a dedicated CD player or a multi-format machine with good CD playback quality. If you'd like to experience high-resolution multichannel music (and have a multichannel playback system), select a player with those capabilities.

If your system is based on an A/V controller or A/V receiver rather than an analog preamplifier, digital-to-analog (D/A) conversion can be performed in the controller or receiver rather than in the player. You simply connect a digital cable between the player and controller or receiver, bypassing the player's digital-to-analog conversion circuitry. This technique replaces the D/A conversion in the player with the D/A conversion inherent in all controllers and receivers. The quality of the audio section in the source player (probably a DVD machine) no longer has an effect on the sound quality. Instead, the D/A conversion circuitry in the controller or receiver determines the quality of the sound from digital sources.

What to Listen For

Perhaps more than any other components, digital sources come in the most "flavors." That is, their sonic and musical characteristics vary greatly between brands and models. This variability has its drawbacks ("Which one is *right*?"), but also offers the music lover the chance to select one that best complements her playback system's characteristics and suits her musical tastes. The different types of musical presentations heard in CD players, SACD machines, DVD-A players, transports, and digital processors tend to reflect their designers' musical priorities. If the designer's parts budget—or skill—is limited, certain areas of musical reproduction will be poorer than others. The trick is to find the processor that, *in the context of your system*, excels in the areas you find most important musically.

Selecting a digital source specifically tailored for the rest of your playback system can sometimes counteract some of the playback system's shortcomings. For example, don't choose a bright-sounding CD player for a system that is already on the bright side of reality. Instead, you may want to select a player whose main attribute is a smooth, unfatiguing treble. Each digital product has its particular strengths and weaknesses. Only by careful auditioning—preferably in your own system—can you choose the product best for you.

To illustrate this, I've invented two hypothetical listeners—each with different systems and tastes—and two hypothetical CD players. I've used a CD player in the example, but CD transports, digital processors, SACD, or DVD-A machines could be easily substituted. Although the following discussion could apply to all audio components, it is particularly true of digital components. Not only are there wide variations in sonic characteristics between processors, but a poor-sounding digital processor at the front-end of a superb system will ruin the overall performance.

Listener A likes classical music, particularly early music, Baroque, and choral performances. She rarely listens to full-scale orchestral works, and never plays rock, jazz, or pop. Her system uses inexpensive solid-state electronics and fairly bright loudspeakers; the combination gives her a detailed, forward, and somewhat aggressive treble.

Listener B wouldn't know a cello from a clarinet, preferring instead electric blues, rock, and pop. He likes to feel the power of a kickdrum and bass guitar working together to drive the rhythm. His system is a little soft in the treble, and not as dynamic as he'd like.

Now, let's look at the sonic differences between two inexpensive and similarly priced CD players and see how each would—or wouldn't—fit in the two systems.

CD player #1 has terrific bass: tight, deep, driving, and rhythmically exciting. Unfortunately, its treble is a little etched, grainy, and overly prominent. CD player #2's best characteristics are its sweet, silky-smooth treble. The player has a complete lack of hardness, grain, etch, and fatigue. Its weakness, however, is a soft bass and limited dynamics. It doesn't have a driving punch and dynamic impact on drums compared to CD player #1.

I think you can guess which player would be best for each system and listener. CD player #1 would only exacerbate the brightness Listener A's system already has. Moreover, the additional grain would be more objectionable on violins and voices. CD player #2, however, would tend to soften the treble presentation in Listener A's system, providing much-needed relief from its relentless treble. Moreover, the sonic qualities of CD player #1—dynamics and tight bass—are less important musically to Listener A.

Conversely, Listener B would be better off with CD player #1. Not only would player #1's better dynamics and tighter bass better serve the kind of music Listener B prefers, but his system could use a little more sparkle in the treble and punch in the bass.

Which CD player is "better"? Ask Listener A after she's auditioned both products in her system; she'll think player #2 is greatly superior, and wonder how anyone could like player #1. But Listener B will find her choice lacking rhythmic power, treble detail, and dynamic impact. To him, there's no comparison: player #1 is the better product.

Though exaggerated for clarity, this example shows how personal taste, musical preference, and system matching can greatly influence which digital products are best for you. The only way to make the right purchasing decision is to *audition the products for yourself*. Use product reviews in magazines to narrow your choice of what to audition. Read reviewers' descriptions of a particular product and see if the type of sonic presentation described is what you're looking for. But don't buy a product solely on the basis of a product review—a reviewer's system and musical tastes may be very different from yours. You could be Listener A and be reading a review written by someone with Listener B's system and tastes.

Use reviews as guides in pointing you to products you might want to audition yourself, not as absolute truth. You're going to spend many hours with your decision, so listen carefully before you buy—it's well worth the investment in time. Moreover, the more products you evaluate and the more careful your listening, the sharper your listening skills will become.

It's important to realize that the specific sonic signatures described in the example are much more pronounced at lower price levels. Two "perfect" digital playback devices would sound identical. At the very highest levels of digital playback, the sonic tradeoffs are much less acute—the best products have very few shortcomings, making them ideal for all types of music.

Still, a significant factor in how good any CD player sounds is the designer's technical skill and musical sensitivity. Given the same parts, two designers of different talents will produce two very different-sounding products. Consequently, it's possible to find skillfully designed but inexpensive products that outperform more expensive products from less talented designers.

Higher-priced products are not necessarily better. Don't get stuck in a specific budget and audition products only within a narrow price range. If an inexpensive product has received a rave review from a reviewer you've grown to trust, and the sonic description matches your taste, audition it—you could save yourself a lot of money. If you decide not to buy the product, at least you've added to your listening database, and can compare your impressions with those of the reviewer.

In addition to determining which digital products let you enjoy music more, there are specific sonic attributes you should listen for that contribute to a good-sounding digital front-end. How high a priority you place on each of these characteristics is a matter of personal taste.

In the following sections, I've outlined the musical and sonic qualities I look for in digital playback.

The first quality I listen for in characterizing how a digital component sounds is its overall perspective. Is it laid-back, smooth, and unaggressive? Or is it forward, bright, and "in my face"? Does the product make me want to "lean into" the music and "open my ears" wider to hear the music's subtlety? Or do my ears tense up and try to shut out some of the sound? Am I relaxed or agitated?

A digital product's overall perspective is a fundamental characteristic that defines that product's ability to provide long-term musical satisfaction. If you feel assaulted by the music, you'll tend to listen less often and for shorter sessions. If the product's fundamental musical perspective is flawed, it doesn't matter what else it does right.

Key words in product reviews that describe an easy-to-listen-to digital product include *ease*, *smooth*, *laid-back*, *sweet*, *polite*, and *unaggressive*. Descriptions of *bright*, *vivid*, *etched*, *forward*, *aggressive*, *analytical*, *immediate*, and *incisive* all point toward the opposite type of presentation.

There is a fundamental conflict between these extremes of presentation. Processors that are smooth, laid-back, and polite may not actively offend, but they often lack detail and resolution. An absence of aggressiveness is often achieved at the expense of obscuring low-level musical information. This missing musical information could be the inner detail in an instrument's timbre that makes the instrument sound more lifelike. It could be the sharp transient attack of percussion instruments; a slight rounding of the attack gives the impression of smoothness but doesn't accurately convey the sound's dynamic structure. Consequently, very smooth-sounding digital products often have lower resolution than more forward ones.

The other extreme is the digital product that is "ruthlessly revealing" of the music's every detail. Rather than smoothing transients, these products hype them. In a side-by-side comparison, a ruthlessly revealing product will appear to present much more detail and musical information. It will sound more upbeat and exciting, and will appeal to some listeners. Such a presentation, however, quickly becomes fatiguing. The listener feels a sense of relief when the music is turned down—or off. The worst thing a product can do is make you want to turn down the volume, or stop listening altogether.

Digital reproduction also has a tendency to homogenize individual instruments within the soundstage. This tendency to blur the distinctions between individual instruments occurs on two levels: the instruments' unique timbral signatures and specific locations within the soundstage.

On the first level, digital products can overlay music with a common synthetic character that diffuses the unique textures of different instruments and buries the subtle tonal differences between them—the music sounds as if it is being played by one big instrument rather than many individual ones. Instead of separate and distinct objects (instruments and voices) hanging in three-dimensional space, the listener hears a synthetic continuum of sound. There is a "sameness" to instrumental textures that prevents their individual characteristics from being heard.

The second way in which digital playback can diffuse the separateness of individual instruments is by presenting images as flat "cardboard cutouts" pasted on top of each other. The instruments aren't surrounded by an envelope of air and space, the soundstage is flat and congested, and you can't clearly hear where one image ends and the next begins. Good digital playback should present a collection of individual images hanging in three-dimensional space, with the unique tonal colors of each instrument intact and a sense of space and air between the instrumental images. This is easy for analog to accomplish, but quite difficult for digital. A recording with an excellent portrayal of timbre and space will help you identify which digital products preserve these characteristics.

Beyond these specifics, a good question to ask yourself is, "How long can I listen without wanting to turn the music down—or off?" Conversely, the desire—or even compulsion—to bring out CD after CD is the sign of a good digital front-end. Some components just won't let you turn off your system; others make you want to do something else.

This ability to musically engage the listener is the essence of high-end audio. It should be the highest criterion when judging digital front-ends.

6

Turntables, Tonearms, and Cartridges

It may come as a surprise to learn that sales of turntables, phono cartridges, and LP records have been steadily increasing since the turn of the millennium. But why would a primitive, early twentieth-century analog technology find growing support in the age of internet downloads and portable digital music players that can store thousands of hours of instantly accessible music?

The answer has several parts. First, the LP record, when played correctly, has a warmth and musicality unmatched by the CD. Some would even argue that the LP is sonically superior to high-resolution digital formats such as DVD-Audio and SACD. Second, the act of setting up and tweaking a turntable, along with the ritual of putting a large black disc on a slowly rotating platter, holds enormous appeal for some music lovers. As one turntable distributor remarked to me "Playing LPs is the barbeque of hi-fi; it's as much about the process as the result." Third, many young people who never knew the turntable and vinyl records as children are now embracing vinyl for its retro-cool factor. In the process, they've discovered the musical pleasure the LP can deliver. Finally, putting an LP on a turntable is an act that signifies a single-minded dedication to focusing on the music. When playing a record you sit down in the listening chair, often with full-sized liner notes and cover art, and with no remote control to skip tracks. The process makes a statement that you are about to give the music your full attention for an entire LP side (at a minimum). Compare that experience with listening to music from a portable music player while engaging in some other activity—one's attention span need be no longer than the time it takes to hit the track-skip button—not to mention the lack of album art and liner notes that are often an important extension of the artist's expression. Spinning an LP represents a return of listening to a central activity, and a rejection of modern society's relegation of music to aural wallpaper.

Returning for a moment to the sound of LP vs. CD, it's worth noting that at the highest levels of music playback, there's not much of a debate; LP is the clear winner. It's interesting to see the number of amplifier and loudspeaker manufacturers at hi-fi shows using a turntable and vinyl records to demonstrate their exotic new products. They want to show their new products in the best light and turn to the LP to deliver the highest quality of sound. When done right, LP playback has an openness, transparency, dynamic expression, and musicality not matched by CD. There's just a fundamental musical rightness to a pure analog source (one that has never been digitized) that seems to better convey the music's expression.

This isn't to say that LPs are perfect. They suffer from a variety of distortions such as mistracking, ticks and pops, speed instability, surface noise, cartridge frequency-response variations, inner-groove distortion, wear, and susceptibility to damage. But for many listeners, these problems are less musically objectionable than the distortion imposed by digitally encoding and decoding an audio signal. (It's been said that turning digits back into music is akin to trying to turn hamburger back into steak.) Some listeners can hear past the LP's flaws and enjoy the medium's overall musicality. Other listeners can't stand the ritual of handling and cleaning records—not to mention keeping the turntable properly adjusted—and think CD is just fine. I think of it this way: LP's distortions are apparent, but separate from the music; digital's distortions are woven into the music's fabric. Consequently, analog's distortions are easier to ignore. If you're inclined to think CD is without fault, and you've never heard a properly played LP, give yourself a treat and visit a specialty audio dealer with a high-end turntable. Listen to what vinyl can do before you write off the possibility of owning a high-quality turntable.

With that background, let's look at the components that make up an LP playback system.

The long-playing (LP) record-playback system is a combination of a turntable, tonearm, and phono cartridge that converts the mechanical information encoded on vinyl records into an electrical signal that can be amplified by the rest of your playback system. The turntable spins the record, the tonearm holds the cartridge in place, and the cartridge converts the wiggles in the record's groove into an electrical signal. Each of these elements, and how they interact with each other, plays a pivotal role in getting good sound from your system. An LP front-end can be as simple as a $350 turntable/arm/cartridge combination (Fig.6-1) that you take out of the box and plug into your system, or an elaborate rig that literally takes several days to set up and tune (Fig.6-2).

Fig.6-1 Turntables can be inexpensive and simple to set up and use. (Courtesy Pro-Ject and Sumiko)

Fig.6-2 Some turntables are elaborate, with a sophisticated suspension system. (Courtesy Basis Audio)

The Turntable

It's easy to think of the turntable as having a minor role in a playback system's sound quality. After all, the turntable only spins the record and holds the tonearm; how much sonic influence could it have?

The answer is surprising: a high-quality turntable is absolutely essential to getting the best performance from the rest of your system. A good turntable presents a solid, vibration-free platform for the record and tonearm, allowing the cartridge to recover the maximum amount of information in the grooves while minimizing interference with the audio signal.

The turntable is composed of a base, platter, platter bearing, plinth, drive system, and often a sub-chassis. The *base* is the turntable's main structure; it holds all the components, and is usually finished in black or natural wood. The *platter* is the heavy disc that supports the record; it rests on the *bearing assembly*. The *plinth* is the top of the turntable beneath the platter. The *drive system* conveys the motor's rotation to the platter. Some turntables have a *sub-chassis* suspended within the base on which the platter and tonearm are mounted. Every turntable will also have an *armboard* for mounting a tonearm. Many turntables have no base or plinth, instead suspending the sub-chassis in open air.

Let's look at how each of these components is assembled into the modern turntable.

Chapter 6

The Base and Plinth

A turntable's base and plinth play important roles in sound quality. The base must be a rigid, vibration-resistant structure on which the other turntable components can be mounted. If the base is flimsy, it will vibrate and transmit that vibration to the platter and tonearm, degrading the sound.

A turntable system can be set vibrating by four forces: 1) acoustic energy impinging on the turntable (called *feedback*); 2) structure-borne vibration traveling through the turntable stand (primarily when the stand is located on a suspended floor); 3) the turntable's mechanical systems, such as the platter bearing and motor vibration; and 4) the motion imparted to the tonearm by groove modulations.

These sources of vibration create relative motion between the stylus and the cartridge. Because the cartridge can't distinguish between groove modulation and turntable resonance, this vibration is converted to an electrical signal and amplified by your system. This is why turntable designers go to elaborate measures to reduce vibration.

Let's first take the case of acoustic energy impinging on the turntable. If the base and plinth aren't rigid, they're more likely to be set in motion by sound striking the turntable. In extreme cases, the loudspeakers and turntable create an *acoustic feedback loop* in which sound from the loudspeakers is converted into an electrical signal in the cartridge through vibration, which is amplified by the loudspeakers, which causes even more feedback to be produced by the cartridge, and so forth. This acoustic feedback can muddy the music, or even make it impossible to play records at a moderately high playback level. You can hear this phenomenon by putting the stylus on the lead-in groove without the record spinning, then gradually turning up the volume. You'll start to hear a howling sound as the acoustic feedback loop grows strong enough to feed on itself, and a "runaway" condition develops in which the sound keeps getting louder. If you try this, keep your hand on the volume control and be ready to turn down the volume as soon as you hear the howl—if you don't turn down the volume *immediately*, the system could be damaged. The more vibration-resistant the turntable, the less severe this phenomenon.

So much for the problem of acoustic energy putting the turntable in motion. Now let's look at how turntable design addresses the problems of structure-borne vibration. Vibration entering the turntable through the stand or rack can be greatly reduced by mechanically isolating the turntable's key components (platter, armboard, and tonearm/cartridge) with a suspension system—the sprung turntable.

Sprung and Unsprung Turntables

Most turntables are sprung, meaning that the platter and armboard are mounted on a sub-chassis that floats within the base on springs. The terms suspended and floating describe the same construction.

Sprung turntables can be one of three designs. In one method, the sub-chassis sits on springs attached to the base bottom. In the second method, the sub-chassis

66

hangs down from the plinth on springs. (Fig.6-3 shows the latter of these two techniques.) The third technique dispenses with the base entirely and hangs the sub-chassis in open air on pillars (Fig.6-2, shown earlier). The turntable is suspended on the four pillars at each corner of the turntable.

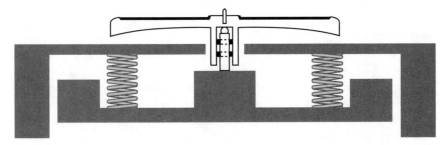

Fig.6-3 The subchassis can be hung from the plinth. (Courtesy Audio Advisor and SOTA Industries.)

Whichever technique is used, the goal of all sprung designs is to isolate the platter and tonearm from external vibration. Any vibration picked up by the supports on which the turntable rests won't be transferred as effectively to the platter and tonearm. The primary sources of structure-borne vibration are passing trucks, footfalls, air conditioners, and motors attached to the building. Structure-borne vibration is much less of a problem in a single-family home with a concrete floor than in an apartment building or frame house with suspended floors.

The Platter and Bearing Assembly

The platter not only provides a support for the record, it also plays two other important roles: as a flywheel, to smooth the rotation; and as a "sink," to draw vibration from the record. Many platters are very heavy (up to 30 pounds), with most of their mass concentrated toward the outside edge to increase their moment of inertia. This high mass also counters bearing friction and stylus drag. Irregular (non-linear) bearing friction can create rapid irregularities in the platter's speed, which frequency-modulates the recovered audio signal. Massive platters greatly reduce the audible effects of bearing friction.

Most platters are made of a single substance such as acrylic, stamped metal (in cheaper turntables), cast and machined aluminum (in better turntables), or exotic materials such as ceramic compounds. The platter sometimes has a hollowed-out ring around the outer edge that is filled with a heavy material to increase the platter's mass, or is loaded with a damping substance to make the platter more inert and resistant to vibration.

These techniques also attempt to make the platter act as a "sink" for record vibration. When the record is clamped to the platter, any record vibration will be transferred to the platter. The platter's material and geometry are thus important design considerations: The platter should have no resonant peaks within the audioband. Some platters use constrained-layer damping and combinations of different materials to provide the ideal sink for record vibration.

Chapter 6

Because the platter spins on a stationary object (the rest of the turntable), there must be a bearing surface between the two. With one technique, the bearing is mounted at the end of a shaft to which the platter is attached. This shaft—often made of stainless steel—extends down a hollow column in the base. The shaft has some form of bearing on the end, either a chrome-hardened steel ball, tungsten-carbide, Zirconium, ceramic, or even a very hard jewel such as sapphire. The bearing often sits in a well of lubricant.

A second technique puts the bearing surface on top of a stationary shaft, with the bearing surface between the platter and shaft.

Whichever technique is used, the bearing must provide smooth and quiet rotation of the platter. Any noise or vibration created by the bearing will be transmitted directly to the platter. Turntable bearings are machined to very close tolerances, and are often highly polished to achieve a smooth surface.

A bearing that suffers none of the traditional mechanical problems is the *air bearing*. The platter rides on a cushion of air rather than on a mechanical bearing. A pump forces compressed air into a very tiny gap between the platter and an adjacent surface. This air pressure pushes the platter up slightly so that the platter literally floats on air. Air-bearing turntables are, however, very expensive and can be difficult to set up. A recent development in turntables is the magnetic bearing, in which the platter floats on a magnetic field. This technique has the advantage of completely isolating the platter, but is enormously expensive to implement.

Platter Mats and Record Clamps

Platter mats are designed to minimize record vibration, and then to absorb what vibration remains. Designers of soft mats suggest that an absorbent felt mat works better in drawing vibration away from the record. Designers of hard mats contend that a stiffer mat material better couples the record to the platter. Finally, some turntable manufacturers discourage using any mat at all, believing that their platter design provides the best sink for record vibration.

Record clamps couple the record to the platter so that the record isn't allowed to vibrate freely (Fig.6-4). The platter acts as a vibration sink, draining vibration from the record. By more intimately coupling the record to the platter, the record clamp improves the sound.

Record clamps come in three varieties. First, the clamp can simply be a heavy weight put over the spindle. The clamp's weight squeezes the record between clamp and platter. Other clamps have a screw mechanism that threads down onto the spindle. The third type is the "reflex" clamp, in which a locking mechanism pushes the clamp down onto the record. Which type works best should be decided by your listening, the turntable manufacturer's recommendations, or your local dealer's suggestion. Note that very heavy clamps can put a strain on some sprung turntables, compressing the springs (or expanding them if the sub-chassis is hung from the plinth). Some form of record clamping is, however, a must for any high-end turntable.

Fig.6-4 A record clamp creates a tight coupling between the record and platter and is essential to good sound. (Courtesy Music Direct)

An extra measure of clamping takes the form of a heavy metal ring whose inner diameter is very slightly larger than an LP. When placed over the LP on the platter, this ring couples the LP's outer edges to the platter. These rings are used in conjunction with, rather than in lieu of, a spindle clamp.

The Drive System

A turntable's drive system transfers the motor's rotation to the platter. Virtually all high-end turntables currently made are belt-driven; the platter is spun by a rubber belt or silk thread stretched around the motor pulley and outer rim of the platter (Fig.6-5).

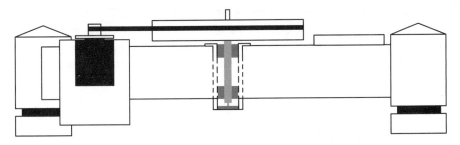

Fig.6-5 Virtually all high-end turntables are belt-driven.

Mass-market mid-fi turntables, when they were still made, were usually *direct-drive*; i.e., the motor is connected directly to the platter. The motor's spindle is often the spindle over which you place the record. Direct drive was sold to the public as superior to belt drive—there are no belts to stretch and wear, and a direct-drive motor can be electronically controlled to maintain precise speed and have low wow and flutter. Indeed, the wow and flutter specifications of a direct-drive turntable are generally better than those of a belt-drive turntable.

But, without question, belt-drive turntables sound far better than direct-drive units. Rather than directly coupling the motor's vibration to the platter as in direct-drive, the drive belt acts as a buffer to decouple the platter from the motor. Motor noise is isolated from the platter, resulting in quieter operation than is possible from a direct-drive turntable. Belt drive also makes it easier to suspend the platter and drive system on a sub-chassis.

No elaborate speed controls are used on belt-drive turntables; the motor just sits there spinning at a fairly high speed (as fast as 1000rpm). This high-speed rotation is coupled to the large platter with a small pulley, resulting in 33-1/3 rpm rotation of the platter.

The drive motor can be a source of turntable vibration. As the motor spins, it produces vibration that can be transferred to the other components in the turntable, producing a low-frequency rumble. Even if you don't hear rumble directly, motor vibration can still degrade the sound. The motor assemblies of some turntables are completely separate from the base and encased in damping material. Other designs mount the motor to the sub-chassis, and isolate its vibration from the other turntable components.

The Tonearm

The tonearm's job is to hold the cartridge over the record and keep the stylus in the groove. We want the tonearm to be an immovable support for the cartridge, yet also be light enough to follow the inward path of the groove, track the up-and-down motions of record warps, and follow any record eccentricity caused by an offset center hole—all without causing undue wear on the delicate grooves themselves. As we'll see, this is a challenging job.

Tonearms come in two varieties: *pivoted* and *tangential-tracking*. A pivoted tonearm (Fig.6-6) allows the cartridge end of the arm to traverse the record in an arc while maintaining a fixed pivot point. A tangential tonearm (also called a *linear tracking* tonearm, shown in Fig.6-2 earlier) moves the entire tonearm and bearing in relation to the record.

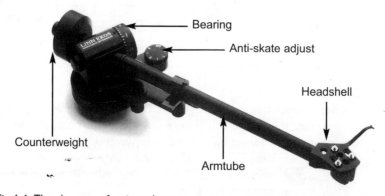

Bearing

Anti-skate adjust

Headshell

Counterweight

Armtube

Fig. 6-6 The elements of a pivoted tonearm. (Courtesy Linn Products)

Turntables, Tonearms, and Cartridges

Let's take a closer look at the pivoted tonearm, by far the most popular type of arm. Its major components are, from back to front, the counterweight, bearing, armtube, and headshell. (These elements are shown from left to right in the photograph of Fig.6-6.) The *counterweight* counteracts the weight of the armtube and cartridge; its weight and position determine the downward force of the stylus in the groove. The *bearing* provides a pivot point for the arm, in both the vertical and horizontal planes. The *armtube* extends the cartridge position away from the pivot point to an optimum position over the record. The *headshell* is attached to the end of the armtube and provides a platform for mounting the cartridge to the armtube. The small, flat disc near the bearing in Fig.6-6 sets the anti-skating compensation.

A tonearm's bearing is an important aspect of its design. The bearing should provide very low friction and not impede the arm's movement. If the bearing is sticky, the stylus will be forced against the groove wall, causing distortion and record wear. Loosening the bearings reduces friction, but can cause the bearings to "chatter" as the tonearm is rattled by the motion of the stylus in the groove, or by other sources of tonearm vibration. Remember, any movement of the tonearm in relation to the stylus in the groove is interpreted by the cartridge as groove modulation, and is converted into an electrical signal that appears at the cartridge output along with the musical signal. Tightening the bearings decreases chatter but also increases friction. Tonearm designers must balance these tradeoffs.

When playing a record, the tonearm is pulled toward the center of the record, a phenomenon called *skating*. Skating is a force acting on the stylus that must be compensated for by applying an equal but opposite force on the cartridge. This compensation, called *anti-skating*, counteracts the skating force caused by tonearm offset. Anti-skating allows the stylus to maintain equal contact pressure with both sides of the groove, and prevents the cartridge's cantilever from being displaced from its center position in the cartridge. Anti-skating can be generated by springs, weights with pulleys, or mechanical linkages.

The Phono Cartridge

The phono cartridge has the job of converting the modulations of the record groove into an electrical signal. Because the cartridge changes one form of energy into another (mechanical into electrical), the cartridge is called a *transducer*. There's one other transducer in your system—the loudspeakers at the other end of the playback chain.

A phono cartridge consists of the cartridge body, stylus, cantilever, and generator system. The *body* is the housing that surrounds the cartridge, and comprises the entire surface area. The stylus is a diamond point attached to the cantilever (the tiny shaft that extends from the bottom of the cartridge body). The stylus is moved back and forth and up and down by modulations in the record groove. This modulation is transferred by the cantilever to the *generator system*, the part of the cartridge where motion is converted into an electrical signal.

Chapter 6

Moving-Magnet and Moving-Coil Cartridges

Cartridges are classified by their principle of operation: moving-magnet or moving-coil. In a moving-magnet cartridge, tiny magnets attached to the cantilever move in relation to stationary coils in the cartridge body. The movement of the magnetic field through the coils induces a voltage (the audio signal) across the coils. (Fig.6-7 shows the essential elements of a moving-magnet cartridge.)

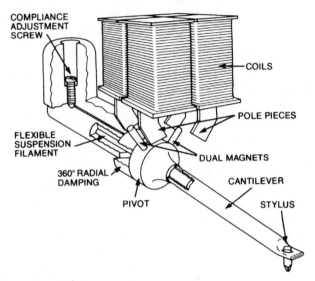

Fig. 6-7 Moving-magnet cartridge construction. (Courtesy Audio Technica)

A *moving-coil* cartridge works on exactly the same physical principles, but the magnets are stationary and the coils move. A moving-coil cartridge generally has much less moving mass than a moving-magnet cartridge. Consequently, a moving-coil cartridge can generally track better than a moving-magnet type, and also have better transient response. With less mass to put into motion (and less mass to continue moving after the motivating force has stopped), moving-coil cartridges can better follow transient signals in the record. Because of their construction, moving-coil cartridges generally don't have user-replaceable styli; you must return the cartridge to the manufacturer.

Cartridge output voltage varies greatly between moving-magnet and moving-coil operation. A moving-magnet's output ranges from 2mV (two thousandths of a volt) to about 8mV; a moving-coil cartridge's output is typically between 0.15mV and 2.5mV. Although moving-coil cartridges generally have lower output voltage than moving-magnet types, some so-called "high-output" moving-coils have higher output voltage than some moving-magnet cartridges.

This wide range of cartridge output voltage requires that the phono preamplifier's gain be matched to the cartridge's output voltage. The lower the cartridge output voltage, the higher the gain needed to bring the phono signal to line level. (A full discussion of matching cartridge output voltage to phono-stage gain is included in Chapter 7.)

The Stylus and Cantilever

Styli (the plural of stylus) come in a variety of shapes, the simplest and least expensive of which is the conical or spherical tip. The conical stylus is a tiny piece of diamond polished into a cone shape. An elliptical stylus has an oval cross section, with two flattish faces. Because this shape more closely approximates the shape of the cutting stylus, it results in lower tracking distortion.

Keeping your stylus clean is of paramount importance: The stylus should be cleaned before every record side. A speck of dust or dirt is like a boulder attached to the stylus, grinding away at the groove walls. An appreciation of the enormous pressure a stylus imposes on the groove further highlights the need for a clean stylus. For example, a tracking force of 1.4 grams applied to a typical stylus contact area results in a pressure of nearly four tons per square inch. This pressure is enough to momentarily melt the outer layer of the groove wall. It's easy to see how stylus motion through the groove is much smoother with a clean stylus, and produces much less record wear. A clean stylus sounds better, too.

A stylus should be cleaned with a back-to-front motion so that the brush follows the record's motion. Some manufacturers recommend that no cleaning fluid be used; others suggest that a fluid is essential to removing accumulated dirt. There is also debate over the best type of brush. Some have short, stiff bristles, while other cleaners resemble nail-polish brushes. Your best bet is to follow the cartridge manufacturer's cleaning instructions. And don't clean the stylus with the tonearm locked in place; you could apply too much force and damage the stylus.

With good maintenance, a stylus should last for about 1000 hours of use. It's a good idea to have the stylus examined microscopically after about 500 hours, then again at 800 hours to check for irregular wear that could damage records.

Because the cantilever transfers stylus motion to the generator, its construction is extremely important. Cantilevers are designed to be very light, rigid, and nonresonant. The lower the cantilever mass, the better the cartridge's trackability, all other factors being equal. To obtain stiffness with low mass, exotic materials are often used in cantilever design, including boron, diamond, beryllium, titanium, ceramic, ruby, and sapphire. Cantilevers are often hollow to reduce their mass, and are sometimes filled with a resonance-damping material.

The cantilever is mounted in a *compliance* inside the cartridge body at the end opposite to that which bears the stylus. The compliance allows the cantilever to move, yet keeps it in position. Because this compliance is stiff when the cartridge is new, it takes many hours of use for the cartridge to break-in and sound its best. It isn't unusual for a cartridge to continue to sound better after 100 hours of use.

Chapter 6

How to Choose an LP Playback System

Because there are many more variables to account for in LP playback, how much of your audio budget you should spend on this part of your system is a more complicated decision than setting a budget for, say, a power amplifier.

Let's look at two hypothetical audiophiles, one of whom should spend much more on a turntable, tonearm, and cartridge than the other audiophile.

Our first audiophile has a huge record collection that represents a lifetime of collecting music. Her record collection is a treasure trove of intimately known music that she plays daily. Conversely, she has very few CDs, buying them only when her favorite music isn't available on vinyl. She much prefers the sound of LPs, and doesn't mind the greater effort required by LPs: record and stylus cleaning, turning over the record, lack of random access.

The second audiophile's record assortment represents a small percentage of his music collection—most of his favorite music is on CD. His LP listening time is a fraction of the time spent listening to CD. He likes the convenience of CD, and can happily live with the sound of his excellent CD player or universal-disc machine.

The first audiophile should commit a significant portion of her overall system budget to a top-notch LP front end—perhaps 40%. The second will want to spend much less—say, 10 to 15%—and put the savings into the components he spends more time listening to.

A decent entry-level turntable with integral tonearm runs about $300, with an appropriate cartridge adding from $30 to $150 to the price. A mid-level turntable and arm costs $800 to $1500; the cartridge price range for this level of turntable is from $200 to $700.

There are roughly two quality and price levels above the $2000 mark. The first is occupied by a wide selection of turntables and arms costing between $3000 and $6000. At this price, you can achieve very nearly state-of-the-art performance. Plan to spend at least $1000 for a phono cartridge appropriate for these turntables.

The next price level is established by turntables costing between $10,000 and $30,000. A topflight phono cartridge can add as much as $5000 to the price. You should know that a super-stratified class of turntables exists, with some models carrying six-figure price tags. These systems are characterized by extraordinary and elaborate construction techniques.

Once you've decided on an LP front-end budget, allocate a percentage of that budget to the turntable, tonearm, and cartridge. Many lower-priced turntables come with an arm—or an arm and cartridge—already fitted and included in the price. At the other end of the scale, the megabuck turntables also include a tonearm. In the middle range, you should expect to spend about 50% of your budget on the turntable, 25% on a tonearm, and 25% on a cartridge. These aren't hard figures, but an approximation of how your LP front-end budget should be allocated. As usual, a local audio retailer with whom you've established a relationship is your best source of advice on assembling the best LP playback system for your budget.

74

Turntables, Tonearms, and Cartridges

An item of utmost importance in achieving good sound from your records is a good turntable stand or equipment rack, particularly if the turntable has a less than adequate suspension. I cannot overstate how vital a solid, vibration-resistant stand is in getting the most from your analog front end. Save some of your budget for a solid equipment rack or you're wasting your money on a good turntable, arm, and cartridge. (The very best turntables have extraordinary mechanical systems to isolate the turntable from vibration, and thus don't benefit as much from a solid stand. These turntables are rare, and occupy the upper end of the price range.)

What to Listen For

Judging the sonic and musical performance of an LP playback system is more difficult than evaluating any other component. If you want to audition a phono cartridge, for example, you cannot do so without also hearing the turntable, tonearm, and how the cartridge interacts with the rest of the LP playback system. The same situation applies to each of the elements that make up an analog front end; you can never hear them in isolation. Further, how those three components are set up greatly affects the overall performance. Other variables include the turntable stand, where it's located in the room, the phono preamp, and the load the preamp presents to the cartridge. Nonetheless, each turntable, tonearm, and cartridge has its own sonic signature. The better products have less of a sonic signature than lower-quality ones; i.e., they more closely approach sonic neutrality.

A high-quality LP front end is characterized by a lack of rumble (or low-frequency noise), greatly reduced record-surface noise, and the impression that the music emerges from a black background. Low-quality LP front ends tend to add a layer of sonic grunge below the music that imposes a grayish opacity on the sound. When you switch to a high-quality front end, it's like washing a grimy film off of a picture window—the view suddenly becomes more transparent, vivid, and alive.

Even if the LP front end doesn't have any obvious rumble, it can still add this layer of low-level noise to the music. The noise not only adds a murkiness to the sound, it also obscures low-level musical detail. A better turntable and tonearm strip away this film and let you hear much deeper into the music. It can be difficult to identify this layer of noise unless you've heard the same music reproduced on a top-notch front end. Once you've heard a good LP playback system, however, the difference is startling. To use a visual analogy, hearing your records on a good analog front end for the first time is like looking at the stars on a cloudless, moonless night in the country after living in the city all your life. A wealth of subtle musical detail is revealed, and with it, a much greater involvement in the music.

Another important characteristic you'll hear on a superlative system is the impression that the music is made up of individual instruments existing in space. Each instrument will occupy a specific point in space, and be surrounded by a halo of air that keeps it separate from the other instruments. The music sounds as if it is made up of individual elements rather than sounding homogenized, blurred, congested, and confused. There's a special realness, life, and immediacy to records played on a high-quality turntable system.

A related aspect is soundstage transparency—the impression that the musical presentation is crystal-clear rather than slightly opaque. A transparent soundstage lets you hear deep into the concert hall, with instruments toward the rear maintaining their clarity and immediacy, yet still sounding far back in the stage. The ability to "see" deep into the soundstage provides a feeling of a vast expanse of space before you in which the instruments can exist (if, of course, the recording engineers have captured these qualities in the first place). Reverberation decay hangs in space longer, further conveying the impression of space and depth.

Conversely, a lower-quality LP front end will tend to obscure sounds emanating from the stage rear, making them sound undifferentiated and lacking life. The presentation is clouded by an opaque haze that dulls the music's sense of immediacy, prevents you from hearing low-level detail, and tends to shrink your sense of the hall's size.

Other important musical qualities greatly affected by the LP front end are dynamic contrast and transient speed. A top-notch LP replay system has a much wider dynamic expression than a mediocre one; the difference between loud and soft is greater. In addition to having wider dynamic range, musical transients have an increased sense of suddenness, zip, and sharpness of attack. The attack of an acoustic guitar string, for example, is quick, sharp, and vivid. Many mediocre turntables and arms slow these transient signals, making them sound synthetic and lifeless. A good LP front end will also have a coherence that makes the transients sound as if they are all lined up in time with each other. The result is more powerful rhythmic expression.

Just as we want the LP front end to portray the steep attack of a note, we want the note to decay with equal rapidity. A first-rate LP front end is characterized by its ability to clearly articulate each note with a sense of silence between the notes, rather than blurring them together. A good test for an LP front end's transient characteristics is intricate percussion music. Any blurring of the music's dynamic structure—attack and decay—will be immediately obvious as a smearing of the sound, lack of immediacy, and the impression that you're hearing a replica of the instruments rather than the instruments themselves. Hearing live, unamplified musical instruments periodically really sharpens your hearing acuity for judging reproduced sound.

All turntables, tonearms, and cartridges influence the sound's overall perspective and tonal balance. Even high-quality components can have distinctive sonic signatures. Careful matching is therefore required between the turntable, arm, and cartridge to achieve a musical result. Matching a bright, forward cartridge to an arm with the same characteristics, and mounting both on a somewhat aggressive-sounding turntable, could be a recipe for unmusical sound. Those same individual components may, however, be eminently musical when in a mix of components that tend to complement each other.

The very best turntables, tonearms, and cartridges are more sonically neutral than lesser products. System matching is therefore less critical as you go up in quality.

State-of-the-art turntables and tonearms tend to be so neutral that you can put together nearly any combination and get superlative sound.

You should listen for two other aspects of turntable performance: *speed accuracy* and *speed stability*. Speed accuracy is how close the turntable's speed is compared to 33-1/3 rpm. You need to worry about speed accuracy only if you have absolute pitch (the ability to identify a pitch in isolation). Speed instability, however, is easily audible by anyone, and is particularly annoying. Speed stability is how smoothly the platter rotates. Poor speed stability causes *wow and flutter*. Wow is a very slow speed variation that shifts the pitch slowly up and down and is most audible on solo piano with sustained notes. Flutter is a rapid speed fluctuation that almost sounds like tremolo. Together, wow and flutter make the sound unstable and blurred rather than solidly anchored. A good turntable with no obvious flutter can still suffer from speed instability. Instead of hearing flutter overtly, you may hear a reduction in timbrel accuracy—an oboe, for example, will sound less like an oboe and more like an undifferentiated tone.

7

Preamplifiers

The preamplifier is the Grand Central Station of your hi-fi system. It receives signals from source components—turntables, CD players, SACD and DVD-Audio machines, FM tuners, satellite radio receivers, music servers—and allows you to select which of these to send to the power amplifier for listening. In addition to allowing you to switch between sources, the preamplifier performs many other useful functions, such as amplifying the signal from your phono cartridge (in some preamplifiers), adjusting the balance between channels, and allowing you to set the volume level. Fig.2-1, shown earlier, illustrates the stereo preamplifier's role in a music-playback system.

In addition to providing volume control and letting you select which source component you listen to, the preamplifier is a buffer between your source components and power amplifier. That is, the preamplifier acts as an intermediary, taking in signals from source components and conditioning those signals before sending them on to the power amplifier. Source components can easily drive a preamplifier, with the burden of driving a power amplifier through long cables falling on the preamp's shoulders. By buffering the signal, the preamplifier makes life easier for your source components and ensures good technical performance.

The preamplifier is the component you will most often use, touch, and adjust. It also has a large influence on the system's overall sound quality. (Note: preamplifiers are built into, or "integrated" with, integrated amplifiers and receivers, instead of being housed in a separate chassis.)

Types of Preamplifiers

There are many types of preamplifiers, each with different capabilities and functions. Choosing the one best suited to your system requires you to define your needs. Listeners without a turntable, for example, won't need a preamplifier with the ability to amplify the tiny signals from a phono cartridge. Others will need many inputs to accommodate digital recorders, music servers, FM tuners, satellite radio, and some will need multichannel capability. Let's survey the various preamplifiers and define some common preamplifier terms.

Line-Stage Preamplifier: Accepts only *line-level* (low-level) signals, which include every source component except a turntable. Line stages have become much more popular as listeners increasingly rely on digital sources rather than LPs as their main signal source. If you don't have a turntable, you need only a line-stage preamplifier.

Phono Preamplifier (also called a *phono stage*): Takes the very tiny signal from your phono cartridge and amplifies it to line level. It also performs *RIAA equalization* on the signal from the cartridge. RIAA equalization, named after the Recording Industry Association of America, is a bass boost and treble cut that counteract the bass cut and treble boost applied in disc mastering, thus restoring flat response. A phono stage can be an outboard stand-alone unit in its own chassis, or a circuit section within a full-function preamplifier. If you play records, you must have a phono stage, either as a separate component or as part of a full-function preamplifier.

Full-Function Preamplifier: Combines a phono-stage with a line-stage preamplifier in one chassis.

Tubed Preamplifier: A tubed preamplifier uses vacuum tubes to amplify the audio signal.

Solid-State Preamplifier: A solid-state preamplifier uses transistors to amplify the audio signal.

Hybrid Preamplifier: A hybrid preamplifier uses a combination of tubes and transistors.

Audio/Video Controller: A device analogous to a preamplifier that includes video switching, multiple audio channels (typically six), and surround-sound decoding such as Dolby Digital and DTS. (A/V controllers are described in Chapter 10, "Audio for Home Theater and Multichannel Music.")

Multichannel Preamplifier: A preamplifier with multiple audio channels (typically six) for playback of multichannel music. Differs from an A/V controller in that the multichannel preamplifier has no surround decoding, video switching, or other functions for film-soundtrack reproduction.

How to Choose a Preamplifier

Once you've decided on a line-stage, a full-function preamp, A/V controller, or separate line and phono stages, it's time to define your system requirements. The first is the number of inputs you'll need. If you have only a CD player, the four or five line inputs on most preamps are more than enough. But let's say you have a turntable, CD player, FM tuner, music server, satellite radio receiver and you also want to plug into your preamp an iPod or portable music player. You'll need a phono input, five line inputs, and one tape loop—at the minimum. A tape loop is a pair of input and output jacks for recording onto a recording device and receiving a signal from that recording device. When you press the TAPE MONITOR button on your preamplifier, you are routing the signal from the TAPE INPUT jacks to your power amplifier for listening.

Most high-end preamps have few features—and for good reason. First, the less circuitry in the signal path, the purer the signal and the better the sound. Second, the preamp designer can usually put a fixed manufacturing budget into making a preamp that either sounds superb or has lots of features—but not both. Mass-market mid-fi equipment emphasizes vast arrays of features and buttons at the expense of sound quality. Don't be surprised to find very expensive preamps with almost no features; they were designed, first and foremost, for the best musical performance. Most high-end preamps don't even have tone (bass and treble) controls. Not only do tone controls electrically degrade the signal—and therefore the musical performance—but the very idea of changing the signal is antithetical to the values of high-end audio. The signal should be reproduced with the least alteration possible. Tone controls are usually unnecessary in a high-quality system.

Another school of thought, however, holds that a playback system's goal isn't to perfectly reproduce what's on the recording, but to achieve the most enjoyable experience possible. If changing the tonal balance with tone controls enhances the pleasure of listening to music, use them. If you're comfortable with the latter philosophy, be aware that tone controls invariably degrade the preamplifier's sonic quality. (Some preamps with tone controls allow you to switch them out of the circuit when they're not being used, a feature called "tone defeat.")

A similar debate rages over whether or not to include a balance control in a high-end preamp. A balance control lets you adjust the relative levels of the left and right channels. If the recording has slightly more signal in one channel than the other, the center image will appear to shift toward the louder channel, and the sense of soundstage layering may be reduced. A similar problem can occur if the listening room has more absorptive material on one side than the other, pulling the image off-center. A balance control can also center the soundstage if you are sitting to the left or right of the sweet spot. A small adjustment of the balance control can correct these problems. Like tone controls, balance controls can slightly degrade a preamplifier's sonic performance. It isn't unusual to find a $10,000 state-of-the-art preamplifier with no tone or balance controls, and a $199 mass-market receiver with both of these features.

When making a purchasing decision, you should also consider the preamplifier's look and feel. Because the preamplifier is the component you'll interact with the most, ask yourself: Are the controls well laid out? Is the volume control easy to find in the dark? Is using the remote intuitive? Does the preamp have a mute switch? A mute switch is handy when your listening is interrupted, for protecting the rest of your system when disconnecting cables, or for maintaining the same volume level when comparing two other components, such as digital sources or cables.

Tubes vs. Transistors

Of the many components that make up a hi-fi system, the preamplifier is the most likely to use vacuum tubes instead of solid-state devices (transistors). This is because preamplifiers handle only low-level signals, making tubes more practical and affordable there than in power amplifiers. Power-amplifier tubes are large and expensive, run hot, and require expensive replacement. Further, one theory of audio design holds that if the system is to include tubes, they are best employed closest to the signal source—such as in the preamplifier.

Moreover, the qualities that have endeared many music lovers to the magic of tubes are much more affordable when used in a preamplifier. Tubed audio components are more expensive to build and maintain than solid-state units, but that cost differential is much lower in preamplifiers than in power amplifiers. If you want the special qualities of tubes but not their heat, greater expense, and higher maintenance, the tubed preamplifier is the way to go.

Tubes are often claimed to sound sweeter and warmer, and to have a more natural treble. Many solid-state preamps tend to make the treble dry, brittle, metallic, and etched. The result is steely-sounding strings (particularly violins), unnaturally emphasized vocal sibilants (*s* and *sh* sounds), and cymbals that sound like bursts of high-frequency noise rather than a delicate brass-like shimmer. Because these unpleasant artifacts can be introduced by many components (digital sources, power amplifiers, cables, tweeters), a natural-sounding tubed preamplifier can tend to counteract the system's tendency toward these amusical characteristics.

Tubes have several advantages in their favor as audio-amplifying devices. First, the circuitry associated with a vacuum tube—the ancillary parts that make the tube work—is generally much simpler than a circuit using transistors. Second, the distortion tubes produce is much more benign than the distortion created by solid-state electronics. Tube distortion is predominantly second-harmonic, which can actually be pleasant to the ear. Transistors produce upper-order harmonics that harden the sound. Tube proponents point to these facts as evidence of the superiority of tubed preamplifiers.

Be aware, however, that some tubed preamps are intentionally designed to sound very colored. Rather than present the music with the least added effect, tubed preamps often add significant amounts of *euphonic coloration*. This form of distortion can at first be pleasing to the ear, but represents a departure from the original signal. This type of "tubey" coloration is characterized by a soft treble, an overly laid-back and easygoing presentation, lack of detail, and a "syrupy" sound. Many audiophiles, in their attempts to avoid the worst characteristics of solid-state, turn to euphonically colored tubed preamps to make their systems listenable.

It is a far better approach, however, to make sure that each component in the chain is as transparent as possible. If this is achieved, there will be no need for "tubey" preamps. Ideally, the listener shouldn't be aware that she is listening to tubes; instead, she should be aware only of the music. Just as poor solid-state preamps color the sound by adding grain, treble hardness, and etch, the poor tube preamp will often err in the opposite direction, obscuring detail, adding false "bloom," and reducing resolution. Be equally aware of both forms of coloration.

It's a mistake to "fall in love" with either solid-state or tubes for the wrong reasons. The solid-state lover may think he is getting "more detail," and the tube aficionado may fall for the "lush sweetness." Both extremes are to be avoided in the pursuit of a truly musical playback system. The overly "sweet" preamplifier may become uninvolving over time because of its low resolution; the "detailed" and "revealing" solid-state unit may eventually become unmusical for the fatigue it produces in the listener. Not all tube and solid-state preamps can, however, be categorized so neatly. Many tubed preamps are extremely transparent and neutral, having very little effect on the sound.

Nor should you buy a preamp purely because it uses tubes. Certain circuits are better implemented with tubes, others with solid-state devices. There are no magic components or circuit designs that will ensure a product's musicality. Worthy—and unworthy—products have been made from both tubes and transistors. That's why some companies design tubed, solid-state, and hybrid tubed/solid-state preamplifiers. The designer picks the best device for the particular application.

Ideally, both tubed and transistor preamplifiers strive for musical perfection, but approach that goal from different directions. The gross colorations I've described are largely a thing of the past, or relegated to fringe products that appeal to a minority of unsophisticated listeners. Today's reality is that tubed and solid-state designs sound more and more alike every year. The best preamplifier will be a transparent window on the music—no matter what its design.

The best advice is to choose the preamp that has the least effect on the music; you'll get much more musical satisfaction from it for a longer period of time. And remember: The perfect tubed preamp and the perfect solid-state preamp would sound identical.

Balanced and Unbalanced Connections

Some preamplifiers have balanced inputs, balanced outputs, or both. A balanced signal is carried on a three-pin XLR connector rather than the conventional RCA plug. (Balanced and unbalanced connections, described in Chapter 10, can be seen in Fig.7-1.) If you have a balanced source component—usually a CD player—you should consider choosing a preamplifier with a balanced input. Nearly all source components with balanced outputs also have unbalanced outputs. You can use either output, but you may not get the best sound quality unless you use the balanced output option. In addition, not all "balanced" preamplifiers are created equal, as we'll see.

A preamp with unbalanced inputs and a balanced output can accept unbalanced signals but still drive a power amplifier through a balanced interconnect. If your power amplifier has balanced inputs, getting a balanced-output preamplifier is a good idea. You can listen to the system through both balanced and unbalanced lines and decide which sounds better. Some products sound better through balanced connections; others perform best with unbalanced lines.

Although all balanced preamps have XLR jacks, not all balanced preamplifiers are created equal. Two preamps that have balanced inputs and outputs can be very different in how they treat the signal. Most preamps accepting a balanced signal immediately convert it to an unbalanced signal, perform the usual preamplifier functions (provide gain and volume adjustment) on the unbalanced signal, then convert the signal back to balanced just before the main output. A preamp with balanced inputs and outputs, but unbalanced internal topology, often adds two active stages to the signal path: a *differential amplifier* at the input and the *phase splitter* at the output. A differential amplifier converts a balanced signal to unbalanced; a phase splitter converts an unbalanced signal to balanced. These additional circuits inevitably degrade sound quality.

The preferred, but much more expensive, method is to keep the signal balanced throughout the preamplifier. This technique requires double the audio circuitry; each portion of the balanced signal is amplified separately. Moreover, very close tolerances between halves of the balanced signal are required. Although the fully balanced circuit has more active devices in the signal path, the signal isn't subjected to a differential amplifier or a phase splitter.

Audio/Video Controllers and Multichannel Preamplifiers

The increasing popularity of home theater, with its multichannel sound, has enticed many high-end preamp manufacturers to produce models that work in 2-channel stereo systems as well as those incorporating multichannel surround sound for home theater. These products are called A/V (for Audio/Video) preamplifiers or A/V controllers. The latter term is more technically correct, and better describes the product's function.

Although A/V controllers are covered in depth in Chapter 10, a brief introduction here is illustrative. A/V controllers are distinguished by three factors: multichannel sound capability, video switching, and surround decoding. First, an A/V controller will typically have six audio channels instead of two to accommodate the 5.1 channels of audio in the Dolby Digital and DTS surround-sound formats. Second, the A/V model can switch video sources as well as audio ones. Finally, A/V controllers all have some form of surround decoding, such as Dolby Pro Logic, Dolby Digital, DTS, or all three.

A key feature of a high-end A/V controller is the ability to pass analog input signals to the output without converting the signal to digital and then back to analog. This analog bypass feature avoids unnecessary sonic degradation imposed by A/D and D/A conversions.

For those who want multichannel music reproduction without video switching and surround decoding, the multichannel preamplifier is a better choice than an A/V controller. A multichannel preamplifier is simply a conventional preamplifier with six channels rather than two (Fig.7-1).

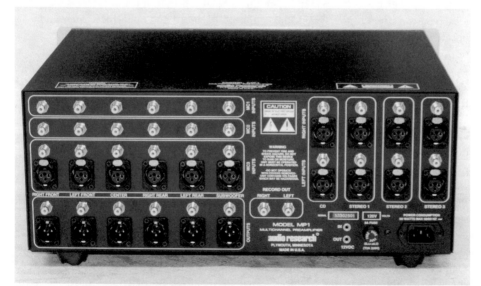

Fig.7-1 A multichannel preamplifier with no home-theater features. (Courtesy Audio Research)

Although it offers no surround decoding, bass management, or other home-theater related functions, it will likely offer better sound quality than a digital controller, for several reasons. First, the chassis contains no digital circuitry, which creates noise that can get into the analog signal. Second, the designer can focus the entire parts budget on making the product sound good without worrying about including all that other complex circuitry. Very few multichannel preamplifiers are available, since they are only for those who will use their systems exclusively for music listening. *PARASOUND P-7*

Phono Preamplifiers

Line-stage preamplifiers are fairly straightforward; they take in line-level signals from the source components, select which source you want to listen to, allow you to adjust the volume, and send that signal to the power amplifier.

A phono preamplifier, however, has a trickier job. The phono preamplifier amplifies the very tiny signal (between a few tens of microvolts and a few millivolts) from the phono cartridge to a line-level (about one volt) signal. This line-level signal can then drive a line-stage preamplifier, just as any other source component would. A phono stage can be an integral part of a full-function preamplifier, an optional board that plugs into some line-stage preamplifiers (Fig.7-2), or an outboard unit in its own chassis (Fig.7-3). Outboard phono stages have no volume controls; they usually feed a line input on a line-stage preamplifier.

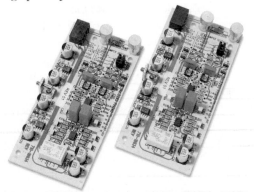

Fig.7-2 A phono preamplifier can be an optional board (or pair of boards) that plugs into a linestage preamplifier. (Courtesy Mark Levinson)

In the days before CD, virtually all preamplifiers included integral phono stages. In many of today's preamplifiers, however, a phono stage is an option (usually about $200–$800) for those listeners who play LPs. This arrangement reduces the preamplifier's price for those needing only a line-stage preamplifier. It's also less expensive to buy a preamplifier with an integral phono stage than buying separate line and phono stages. With one chassis, one power supply, one owner's manual, and one shipping carton, a full-featured preamp is less expensive. These integral phono stages can also offer exceptional performance.

Chapter 7

Fig. 7-3 An outboard phono preamplifier can be a simple box. (Courtesy Pro-Ject and Sumiko)

RIAA Equalization

In addition to amplifying the cartridge's tiny output voltage, the phono stage performs *RIAA equalization* on the signal. RIAA stands for Recording Industry Association of America, the body that standardized the record and playback equalization characteristics. Specifically, phono-stage RIAA equalization boosts the bass and attenuates the treble during playback. This equalization counteracts the bass cut and treble boost applied to the signal when the record was cut. By combining exactly opposite curves in disc mastering and playback, a flat response is achieved. Attenuating bass and boosting treble when the disc is cut allows more signal to be cut into the record groove and increases playing time. Because bass takes up more room in the groove than does treble, reducing the amount of bass in the groove allows the grooves to be spaced closer together. In addition, the treble boost in disc cutting and subsequent treble cut on playback act as a noise-reduction system: attenuating treble on playback also attenuates record-surface noise. RIAA equalization is why you can't plug a line-level source into a preamplifier's phono input. Similarly, a phono cartridge can't be plugged into a preamplifier's line input.

Phono-Stage Gain

The amount of amplification provided by a phono stage (or any other amplifier) is called its *gain*. Gain is specified either in decibels (dB), or as a number expressing the ratio between input and output voltages. Phono stages have much more gain than line stages. Where a line stage may have 10 to 20dB of gain, a phono stage typically amplifies the signal by 35 to 60dB.

The amount of phono-stage gain required depends on the type of phono cartridge driving the phono preamplifier. Phono stages are of two varieties, each named for the type of cartridge with which it is designed to work. The first is the *moving-magnet* phono stage. Moving-magnet phono stages have their gain optimized to work with the relatively high output voltages from moving-magnet cartridges. Moving-magnet cartridges have high output voltages as cartridges go, on the order of two to eight millivolts (2–8mV). Consequently, they need less gain; a moving-magnet phono stage's gain is toward the lower end of the range, typically about 35dB.

Moving-coil cartridges have much lower output voltages due to their different method of generating a signal. Moving-coil output levels range from 0.15mV to 2.5mV. Consequently, they need more amplification (gain) in the phono preamplifier to reach line level than do moving-magnet cartridge signals. Moving-coil phono preamplifiers have about 40–60dB of gain. Note that moving-coil output voltages vary greatly with the cartridge design, with some so-called "high-output" models reaching moving-magnet levels.

Because of this wide variation in cartridge output level, a gain mismatch can occur between the cartridge and phono stage: The phono stage can have either too much or too little gain for a specific cartridge output voltage. If the phono preamp doesn't have enough gain, the volume control must be turned up very high for sufficient playback levels. This raises the noise floor (heard as a loud background hiss), often to the point of becoming objectionable. Conversely, a high-output cartridge can overload a moving-coil phono stage's input circuitry, causing distortion during loud musical passages. This condition is called *input overload*. A high-output cartridge driving a high-gain phono preamplifier can also make the preamplifier's volume control too insensitive. A moderate listening level may be achieved with the volume control barely cracked open; this makes small volume adjustments difficult.

Correctly matching the cartridge output voltage to the phono-stage gain avoids excessive noise and the possibility of input overload. A moving-coil cartridge specified at 0.18mV output needs about 55dB of gain. A typical moving-magnet output of 3mV should drive a phono stage that has about 35dB of gain. Some phono stages and full-function preamps have internal switches that adjust the gain between moving-magnet and moving-coil levels.

High phono-stage gain carries the penalty of increased noise. Although phono stages in general have poorer signal-to-noise specifications than other components, very-high-gain phono preamps can be objectionably noisy. All other factors being equal, the greater the gain, the higher the noise. Select a phono preamp with just enough gain for your cartridge.

Cartridge Loading

Cartridge loading is the impedance and capacitance the phono cartridge "sees" when driving the phono input. Cartridge loading, specified in both impedance and capacitance, has a large effect on how the cartridge sounds, particularly moving-magnet types. Improper loading can cause frequency-response changes and other undesirable conditions. Many preamps allow you to adjust the input impedance and input capacitance to match the phono cartridge by adding resistors and capacitors to the phono stage's input circuit. These adjustments usually require a soldering iron, however, and should be done by your dealer. Some preamps have tiny internal switches to adjust cartridge loading, while others provide front-panel adjustments. One model even allows cartridge loading by remote control while in the listening seat. More discussion of cartridge loading is included in Chapter 6 ("The LP Playback System").

What to Listen For

The preamplifier has a profound effect on the music system's overall performance. Because each of the source signals must go though the preamp, any colorations or unmusical characteristics it imposes will be constantly overlaid on the music. You can have superb source components, a top-notch power amplifier, and excellent loudspeakers, yet still have mediocre sound if the preamp isn't up to the standards set by the rest of your components. The preamplifier can establish the lowest performance level of your system; careful auditioning and wise product selection are crucial to building the best-sounding playback system for your budget.

A preamplifier's price doesn't always indicate its sonic quality; I know of one $1500 model that is musically superior to another preamp selling for nearly $8000. If you do your homework and choose carefully, you'll avoid paying too much for a poor-sounding product.

In addition to the usual listening procedures described in Chapter 4, preamplifiers offer several methods of sonic evaluation not possible with other components. We can therefore more precisely evaluate preamps and choose the best one for the money.

We'll start with the standard listening-evaluation techniques. First, the same musical selection can be played on the same system, alternating between two competing preamps. Be sure to match levels between the two preamps under audition. Listen for the presentation differences described in Chapter 4—particularly clarity, transparency, lack of grain, low-level detail, soundstaging, and a sense of ease.

The most common sonic problems in preamplifiers are a bright and etched treble and a thickening of the soundstage. Many preamps, particularly inexpensive solid-state units, overlay the midrange and treble with a steely hardness. These preamps can give the impression of more musical detail, but quickly become fatiguing. The treble becomes drier, more forward, and etched. Cymbals lose their sheen, instead sounding like bursts of white noise. Vocal sibilants (*s* and *sh* sounds) become objectionably prominent; violins become screechy and thin. The poor-sounding solid-state preamplifier emphasizes the brightness of the strings and diminishes the resonance of the violin's wooden body. Such preamps also reduce the saturation of tonal color, sounding thin and bleached.

The preamplifier that thickens the sound makes the soundstage more opaque. The transparent quality is gone, replaced by a murkiness that obscures low-level detail and reduces resolution. Instruments and voices no longer hang in a transparent, three-dimensional space. Instead, the presentation is thick, confused and congealed, and lacks clarity. Even some expensive models impose these characteristics on the music.

Beware of the preamplifier that tends to make all recordings sound similar in timbre, tonal balance, or spatial presentation. The best preamplifiers let you hear changes in the size of the concert hall, how closely the microphones were positioned, and can clearly resolve the differences between a Steinway piano and a Bösendorfer, among other such subtleties. In addition, listen for how well the preamp resolves the fine inner detail of instrumental timbres.

Compare the preamps under audition to the very best preamp in the store. Listen for the qualities distinguishing the best preamp, and see if those characteristics are in the preamps you're considering. This will not only give you a reference point in selecting a preamplifier for yourself, but will sharpen your listening skills. The more experience you have listening to a variety of products, the better your ability to judge component quality.

A less analytical method is to borrow the preamp from your dealer for a weekend and just listen to it. How much more exciting and involving is the music compared to using your existing preamp? Does the new preamp reveal musical information you hadn't heard before in familiar recordings? How much does the new preamp *compel* you to continue playing music? These are the best indicators of the product's ability to provide long-term musical satisfaction. Trust what your favorite music tells you about the preamplifier.

8

Power and Integrated Amplifiers

The power amplifier—the last component in the signal chain before the loud-speakers—is the workhorse of a hi-fi system. It takes the low-level signal from the preamplifier and converts it to a powerful signal to drive the loudspeakers. It has a low-level input to receive the signal from the preamplifier, and terminals for connecting loudspeaker cables.

Because of the power amplifier's unique function, it differs from other components in size, weight, and use. High-quality power amplifiers are usually large and heavy. Moreover, the power amplifier is the component you don't need to touch and adjust. Power amplifiers are often placed on the floor near the loudspeakers rather than in an equipment rack.

Unlike most of the other components in your system, power amplifiers vary greatly in electrical performance. Consequently, choosing a power amplifier requires careful system matching for electrical compatibility, not just musical compatibility. While any CD player or preamplifier will function in a system (even though it may not be musically ideal), some power amplifiers just won't work well with certain loudspeakers on a technical level. Choosing a power amplifier is therefore a technical, as well as aesthetic, decision that requires careful attention to system matching. We'll discuss these technical factors throughout this chapter.

Let's first survey the types of power amplifiers and define some of the terms associated with them.

Stereo Power Amplifier: A power amplifier with two audio channels (left and right) in one chassis.

Monoblock Power Amplifier: A power amplifier with only one audio channel per chassis. Two monoblocks are required for stereo reproduction.

Multichannel Power Amplifier: An amplifier with, typically, five amplifier channels, for use in a home-theater or multichannel-audio system.

Watt: A unit of electrical power. Power is the ability to do work; in this case, the ability of the amplifier to make a loudspeaker's diaphragm move.

Power Output: The maximum amount of power the amplifier can deliver to the loud-speaker, measured in watts.

Load: In power-amplifier terminology, a load is the loudspeaker(s) the power amplifier must drive.

Tubed: A tubed power amplifier uses vacuum tubes to amplify the audio signal.

Solid-State: A solid-state power amplifier uses transistors to amplify the audio signal.

Chapter 8

Hybrid: A power amplifier combining vacuum tubes and solid-state devices. The input and driver stages are usually tubed, the output stage is usually solid-state.

Class-A: A power amplifier in which the output device or devices (tube or transistor) is operated in such a way that it amplifies the entire musical waveform.

Push-Pull: A power amplifier in which pairs of output devices (tubes or transistors) alternately "push" and "pull" current through the loudspeaker, with one output device amplifying the positive half of the waveform, another output device amplifying the negative half of the waveform. Also called *Class-B* amplification, although not all push-pull amplifiers are Class-B. (Compare with "single-ended.")

Class A/B: A power amplifier that operates in Class A at low power and switches to Class-B operation at higher powers.

Bridging: Converting a stereo power amplifier into a monoblock power amplifier. Some power amplifiers have a rear-panel bridging switch. Also called "strapping" into mono.

Bi-amping: Driving a loudspeaker's midrange and treble units with one amplifier, the woofer with a second amplifier.

Class-D Power Amplifier: An amplifier that operates by switching its output transistors fully on or fully off in a series of pulses. Often erroneously called a "digital" amplifier, but correctly called a "switching" amplifier.

A Survey of Amplifier Types

As you can see from this list, high-performance power amplifiers are available in a wide range of configurations. In fact, high-end amplifiers run the gamut from a minimalist 3Wpc (watts per channel) design to massive 500W behemoth monoblocks. Power amplifiers also cover a huge range of technologies from vacuum tubes to today's high-technology "switching" circuits. You can buy amplifiers with one, two, three, five, or seven amplification channels.

Let's survey the amplifier types before looking at how to choose the right amplifier for your needs.

Monoblock, Stereo, Three-Channel, and Multichannel Amplifiers

The simplest power amplifier is the monoblock. It houses a single amplifier channel in one chassis, with two monoblocks required for stereo reproduction. The familiar stereo amplifier provides two amplification channels, and a multichannel power amplifier offers either five or seven channels. If you plan on using your system only for two-channel music listening, you'll choose a pair of monoblocks or a stereo amplifier.

A recent addition to this list of amplifier configurations is the three-channel amplifier. These units appeal to those who want to convert a two-channel system to multichannel, either to reproduce film soundtracks in a home theater or for multichannel music listening. The three-channel amplifier also makes sense for those whose primary focus is two-channel music with occasional home theater watching—more of the budget can be put into the critical left and right stereo amplification channels, with a less expensive three-channel amplifier handling the center- and surround-channel duties.

The first division of power amplifiers—a stereo unit or a pair of monoblocks—will be decided by your budget. Monoblocks generally *start* at about $2500 per pair. At this price level, a single stereo unit may make more sense; with only one chassis, power cord, and shipping carton, the manufacturer can put more of the manufacturing cost into better parts and performance. I advise against monoblocks if your amplifier budget is less than about $4000. There may be exceptions to this figure, but it nonetheless offers a broad guideline. Many excellent stereo units, for example, cost upward of $6000. A very popular price range for high-quality power amplifiers is $800–$2000, with the $2000 models sometimes offering musical performance close to that of the most expensive amplifiers.

Monoblocks generally perform better than a single stereo unit for several reasons. First, because the two amplifier channels are housed in separate chassis, there is no chance of interaction between channels. Consequently, monoblocks typically have better soundstage performance than stereo units. Second, monoblocks have completely separate power supplies, even down to the power transformers: the left- and right-channel amplifier circuits don't have to share their electrical current source. This gives monoblocks the ability to provide more instantaneous current to the loudspeaker, all other factors being equal. Finally, most manufacturers put their cost-no-object efforts into monoblocks, which are often the flagships of their lines. If you want all-out performance and can afford them, monoblocks are the way to go. A high-quality stereo amplifier is, however, more than sufficient for most systems.

A multichannel audio system can also be based on a single five-channel or seven channel amplifier. The five-channel amp is the more common configuration for two reasons; most multichannel loudspeaker arrays are based on five speakers plus a subwoofer, and squeezing seven amplification channels in a single chassis is a challenge. Some home-theater systems are based on seven channels plus a subwoofer (see Chapter 10 for a full explanation) and require seven amplifier channels. Such a system can be driven by a single seven-channel amplifier or by a five-channel unit augmented with a stereo amp, for example.

Integrated Amplifiers

At the other end of the scale from the monoblock is the integrated amplifier, in which a preamplifier and a power amplifier are combined in the same chassis. Though the power output from integrated amplifiers is generally lower than that from separate power amplifiers, integrateds are much less expensive, and ideal for budget to moderately priced systems. High-quality integrated amplifiers start at about $350, with some models running as high as $9000.

Chapter 8

High-end integrated amplifiers have changed radically in the past few years. Once relegated to low-powered units from European manufacturers with idiosyncratic operation and non-standard connectors, integrated amplifiers have finally come into their own. Leading high-end manufacturers have realized that an integrated amplifier makes sense for many music lovers. The cost and convenience advantages of an integrated amplifier are compelling: integrateds take up less space, are easier to connect, reduce the number of cables in your system, and can even offer the performance of separate components. Now that high-end manufacturers have taken the integrated amplifier seriously, they're putting their best technology and serious design efforts into their integrateds.

Consequently, many manufacturers have enjoyed booming sales of integrated amps in the price range of lower-cost separates that produce about 50–150Wpc (watts per channel) of power. One such product is shown in Fig.8-1.

Fig.8-1 Today's integrated amplifiers offer performance that competes with separate preamplifiers and power amplifiers. (Courtesy Mark Levinson)

Some manufacturers have even included a quality tuner with their integrated amplifier. Not so long ago, the term "high-end stereo receiver" was an oxymoron. Today, however, there's no reason why a receiver designed and built with the dedication given to separate components should offer anything but high-end musical performance.

These newer integrated amplifiers have also overcome one of the limitations of earlier designs: the inability to upgrade just the power amplifier or preamplifier section. Today's integrateds often include preamplifier-out jacks for connecting the integrated to a separate, more powerful amplifier. They also often have power-amplifier input jacks if you want to upgrade the preamplifier section.

When choosing an integrated amplifier, combine the advice in Chapter 7 ("Preamplifiers") with the guidelines in the rest of this chapter. If your budget is under $3000 for amplification, seriously consider one of the new breed of high-end integrated amplifiers rather than separates.

Tubed Power Amplifiers

When transistor amplifiers were introduced in the 1960s, it appeared that the days of the vacuum tube were over. Transistors were smaller, lighter, and cheaper than tubes, ran cooler, and produced more output power. If that wasn't enough, transistor amplifiers didn't need an output transformer, a component that added considerably to the amplifier's size, weight, and cost. All the audio manufacturers of the day scrapped their tubed amplifiers in favor of transistor units virtually overnight.

But many music lovers found the sound of these newfangled "solid-state" amplifiers unlistenable. They likened the sound to that of a pocket transistor radio, only louder. Unfortunately, those perceptive audiophiles couldn't buy a new tubed amplifier after the transistor's introduction.

The modern tubed amplifier was created in 1970 by William Zane Johnson of Audio Research Corporation. As did many music lovers, he found the sound of tubed amplifiers more musical. Johnson's demonstration of a tubed amplifier at a 1970 hi-fi show prompted one show-goer to remark, "You've just set the audio industry back 10 years!" But instead of being a setback, Johnson's amplifier began a renaissance in tubed equipment that is still going strong nearly 50 years later.

The perennial tubes-vs.-transistors debate arises when you're faced with choosing a power amplifier. Tubed units can offer stunning musical performance, but they have their drawbacks. Here are the advantages and disadvantages of tubed power amplifiers. (For the moment, I'll confine my observations to conventional "push-pull" tubed amplifiers, not the exotic single-ended triode varieties described later in this chapter.)

Tubed power amplifiers are more expensive than their similarly powered solid-state counterparts. The cost of tubes, output transformers (output transformers are not needed in solid-state amplifiers), and more extensive power supplies all make owning a tubed power amplifier a more expensive proposition than owning a solid-state unit. Moreover, the tubes will need replacing every few years, further adding to the real cost of ownership.

In terms of bass performance, tubed power amplifiers can't compete with good solid-state units. Tubes have less control in the bass, making the presentation less punchy, taut, and extended. Further, tubed power amplifiers often have limited current delivery into low-impedance loads, making them a poor choice for current-hungry loudspeakers. Tubes also require monthly biasing (small adjustments made with a screwdriver) to maintain top performance. Biasing is very easy, but some music lovers would prefer not to have to think about performing routine maintenance on their music-playback systems.

Power-amp tubes can fail suddenly, sometimes in smoke and (momentary) flames. Such dramatic failure is rare, however; I've used tubed amplifiers for most of my time as a reviewer (since 1989) and have had two tubes fail in that time, both of them uneventfully.

Chapter 8

Finally, there is the possibility of small children or pets burning themselves on hot tubes. Many, but not all, tubed power amplifiers have exposed output tubes. If you have small children, consider a tubed amplifier that is surrounded by a ventilated metal cage.

Given these drawbacks, why would anyone want to own a tubed power amplifier? It's simple: tubes can sound magical. When matched with an appropriate loudspeaker, tubed power amplifiers offer unequaled musicality, in my experience. Even small, moderately priced tubed amplifiers have more than a taste of tube magic.

Many important aspects of music reproduction seem to come naturally to tubed amplifiers. They generally have superb presentation of instrumental timbre, smooth and unfatiguing treble, and spectacular soundstaging. The hard, brittle, edgy midrange and treble presentation of many solid-state amplifiers is contrasted with the purity of timbre and sense of ease conveyed by a good tubed amp. Music has a warmth, ease, and natural musicality when reproduced by many tubed designs. The soundstage has an expansive quality, with a sense of bloom around instrumental images. This isn't to say there aren't good-sounding solid-state power amplifiers, only that tubes seem to more consistently deliver the musical goods. A tubed amplifier's softer bass is often willingly tolerated for its magical midrange, treble, and soundstaging.

As good as tubed amplifiers can sound, solid-state amplifiers have some decided sonic and technical advantages. For example, tubed units are no match for solid-state amplifiers in bass performance. Transistor power amplifiers have tighter, deeper, and much more solid bass than tubed units. The feeling of bass tautness, kick, extension, and power are all better conveyed by solid-state amplifiers, regardless of how good the tubed amplifier is. Speaking technically, solid-state amplifiers can deliver more current to low-impedance loudspeakers, making them a better choice for such loads.

No one but you can decide if a tubed power amplifier is ideal for your system. I strongly suggest, however, that you audition at least one tubed amplifier before making a purchasing decision. You may get hooked.

Single-Ended Triode Amplifiers

So far I've described mainstream power amplifiers that most audiophiles are likely to buy. But a number of variations on the basic design are a significant force in the amplifier marketplace. These include the *single-ended triode* amplifier, the *single-ended solid-state* amplifier, and the *switching,* or *Class-D,* power amplifier.

Single-ended triode amplifiers are an exotic species of power amplifier that has gained a considerable following in the past decade. The single-ended triode (SET) amplifier was the first audio amplifier ever developed, dating back to Lee De Forest's patent of the triode vacuum tube in 1907 and his triode amplifier patent of 1912. SET amplifiers typically produce less than 10Wpc.

You heard right: Large numbers of audiophiles are flocking to replace their modern power amplifiers with 10-watt amplifiers based on 100-year-old technology.

Have the past 100 years of amplifier development been a complete waste of time? A surprising number of music lovers and audio designers think so.

In a single-ended triode amplifier (Fig.8-2), the triode (the simplest of all vacuum tubes) is operated so that it amplifies the entire audio signal. That's what "single-ended" means. Virtually all other power amplifiers are Class-B, meaning that one tube (or transistor) handles the positive half of the musical waveform and a second tube (or transistor) handles the negative half.

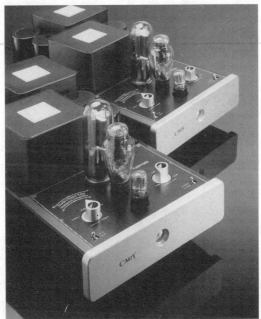

Fig.8-2 The single-ended triode (SET) amplifier must be matched with high-sensitivity loudspeakers. (Courtesy Cary Audio Design)

On the test bench, SET amplifiers have laughably bad technical performance. They typically produce fewer than 10Wpc of output power, and have extremely high distortion—as much as 10% Total Harmonic Distortion (THD) at the amplifier's rated output.

Despite these technical drawbacks, my listening experience with SET amplifiers suggests that this ancient technology has many musical merits. SET amps have a certain presence and immediacy of musical communication that's hard to describe. It's as though the musicians aren't as far removed from here-and-now reality as they are with push-pull amplifiers. SET amps also have a wonderful liquidity and purity of timbre that is completely devoid of grain, hardness, and other artifacts of push-pull amplifiers. When I listen to SET amplifiers (with the right loudspeakers), it's as though the musicians have come alive and are playing in the listening room for me. There's a directness of musical expression that's impossible to put into words, but is immediately understood by anyone who has listened for themselves. You must hear a SET firsthand to know what the fuss is about; no description can convey how they sound.

Chapter 8

When auditioning an SET amplifier, it's easy to be seduced by the midrange. That's because SET amplifiers work best in the midband, and less well at the frequency extremes of bass and treble. If the SET demo is being run for your benefit, be sure to listen to a wide variety of music, not just small-scale music or unaccompanied voice—these will accentuate the SET's strengths and hide its weaknesses.

The importance of matching an SET amplifier to the right loudspeaker cannot be overemphasized. With a low-sensitivity speaker, the SET will produce very little sound, have soft bass, and reproduce almost no dynamic contrast. The ideal loudspeaker for an SET amplifier has high sensitivity (higher than 96dB/1W/1m), high impedance (nominal 8 ohms or higher), and no impedance dips (a minimum impedance of 6 ohms or higher). Such a speaker will produce lots of sound for a small amount of input power, and require very little current.

Single-Ended Solid-State Amplifiers

Single-ended amplifiers aren't confined to those using ancient vacuum-tube technology. Transistors can also be configured to amplify the entire musical waveform. These amplifiers are also called Class-A amplifiers because the mode of operation in which the output device amplifies the entire audio signal is called Class-A. A solid-state, single-ended amplifier is shown in Fig.8-3. Note the large heatsinks required to dissipate the additional heat produced by Class-A operation.

Fig.8-3 A single-ended solid-state power amplifier runs as hot at idle as at full power, requiring large heatsinks to keep it cool. (Courtesy Pass Labs)

Single-ended solid-state amplifiers have better technical performance than single-ended triode amps, with a lower output impedance, more power, and the ability to drive a wider range of loudspeakers. They share many of the benefits of SET amps, particularly the very simple signal path, lack of crossover distortion, and greater linearity. Crossover distortion occurs on each cycle of the waveform when the transistor or tube handling the positive half of the waveform "hands off" the signal to the transistor or tube handling the negative half of the waveform. Although single-ended solid-state amplifiers produce less power than their push-pull counterparts, they generally have much more output power than single-ended tubed units. Nonetheless, it's a mistake to equate single-ended solid-state with single-ended tube amplifiers: there are so many other design variables that single-ended solid-state and single-ended tubed amplifiers should be considered completely different animals.

Class-D ("Switching") Amplifiers

If single-ended triode amplifiers represent a return to fundamental technology, the switching power amplifier may represent the future of audio amplification. Switching amplifiers, also called Class-D amplifiers, have been gaining in popularity due to their small size, low weight, high efficiency, and low cost (Fig.8-4). At the low-end of the audio spectrum, switching amplifiers are becoming ubiquitous in home-theater-in-a-box units. A home-theater-in-a-box may need to power six loudspeakers from a DVD-player-sized chassis—all for a few hundred dollars. Such a unit can output perhaps 300Wpc (50Wpc x 6), yet run cool enough to be placed in a cabinet. In this application, the advantages of a switching amplifier are undeniable. But are switching amplifiers suitable for high-end systems?

Fig.8-4 Class-D amplifiers are small, lightweight, efficient, and produce very little heat. These amps are smaller than a brick and can be held in an outstretched hand. (Courtesy Kharma)

Before tackling that question, let's first look at how a switching amplifier works. In a conventional amplifier (called a *linear amplifier*), the output transistors or tubes amplify a continuously variable analog signal—the musical waveform. In a switching amplifier, the continuously variable analog input signal is converted into a series of on and off pulses. These pulses are fed to the output transistors, which turn the transistors fully on or fully off. When the transistors are turned on, they conduct the DC supply voltage to the loudspeaker. When turned off, no voltage is connected to the loudspeaker. The audio information is contained in the durations of these on-off cycles. The train of pulses amplified by the transistors is smoothed by a filter to recover the musical waveform and remove the switching noise. Because the signal amplitude is contained in the width of the pulses, switching amplifiers are also called pulse-width modulation (PWM) amplifiers.

A switching amplifier operates at very high efficiency, thus the low heat dissipation. Because they run cool, switching amplifiers don't need the large heat sinks found in conventional amplifiers, saving space and cost. Another benefit of this high efficiency is the reduced current demands on the power transformer, which can be smaller, lighter, and cheaper. A Class-A/B amplifier operates at about 40% efficiency (40% of the AC power pulled from the wall socket is converted into power that drives the loudspeakers). By contrast, a Class-D amplifier operates at about 90% (or more) efficiency.

Nonetheless, some successful high-end amplifiers employ switching technology. The field is relatively new, and manufacturers are finding ways to get good sound from switching amplifiers. The technology is in its infancy, suggesting that switching technology may have a future in products other than car stereo and home-theaters-in-a-box.

How to Choose a Power Amplifier

Quiz time: Which stereo system will play louder—one with a 10-watt amplifier or one with a 200-watt amplifier? The answer is that it's impossible to determine without knowing one essential fact: the *sensitivity* of the loudspeaker that the amplifier is driving. A loudspeaker's sensitivity specification indicates how much of the power driving the speaker is converted into sound and how much is wasted as heat. Everyone pays attention to an amplifier's "watts per channel" rating but few consider a loudspeaker's sensitivity. As we'll see, the two specifications are equally important in determining how loudly an audio system will play.

Technically, loudspeaker sensitivity specifies how high a sound-pressure level (SPL) the loudspeaker will produce when driven by a certain power input. A typical sensitivity specification will read "88dB SPL, 1W/1m." This means that the loudspeaker will produce an SPL of 88 decibels (dB) with one watt of input power when measured at a distance of one meter. Although 88dB is a moderate listening volume, a closer look at how power relates to listening level reveals that we need much more than one watt for music playback.

Each 3dB increase in sound-pressure level requires a doubling of amplifier output power. Thus, our loudspeaker with a sensitivity of 88dB at 1W would produce 91dB with 2W, 94dB with 4W, 97dB with 8W, and so on. For this loudspeaker to produce musical peaks of 109dB, we would need an amplifier with 128W of output power.

Now, say we had a loudspeaker rated at 91dB at 1W/1m—only 3dB more sensitive than the first loudspeaker. We can quickly see that we would need only *half* the amplifier power (64W) to produce the same volume of 109dB SPL. A loudspeaker with a sensitivity of 94dB would need just 32W to produce the same volume. The higher-sensitivity speaker simply converts more of the amplifier's power into sound.

To recap, a 3dB increase in loudspeaker sensitivity is equal to doubling the amplifier's output power. A 6dB increase is equal to quadrupling the amplifier's output power, and so on.

This relationship between amplifier power output and loudspeaker sensitivity was inadvertently illustrated in an unusual demonstration nearly 60 years ago. In 1948, loudspeaker pioneer Paul Klipsch conducted a demonstration of live vs. reproduced sound with a symphony orchestra and his Klipschorn loudspeakers. His amplifier power: 5W. The Klipschorns are so sensitive (an astounding 105dB SPL, 1W/1m) that they will produce very high volumes with very little amplifier power. Klipsch was attempting to show that his loudspeakers could closely mimic the tonal quality and loudness of a full symphony orchestra.

The other end of the speaker-sensitivity spectrum was illustrated by a demonstration I attended at the Consumer Electronics Show of an exotic new loudspeaker. During the demo, the music was so quiet that I could barely hear it. I looked at the power amplifiers—300Wpc monsters with large power meters—and was astonished to see that the power meters were nearly constantly pegged at full power. This unusual speaker converted only a minuscule amount of the amplifier's output power into sound.

The importance of loudspeaker sensitivity is also demonstrated by today's 3Wpc single-ended triode amplifiers, which can produce moderately loud listening levels when coupled with high-sensitivity speakers. These examples of huge variations in sound-pressure level and amplifier power illustrate how loudspeaker sensitivity greatly affects how big an amplifier you need. Even a small difference in loudspeaker sensitivity—2dB, say—changes your amplifier power requirements.

In practice, loudspeakers vary in their sensitivities from about 83dB on the low side to perhaps 103dB on the high side, with most falling between 85dB and 94dB. Even the range of 85dB to 94dB represents an eight-fold difference in amplifier power to achieve the same sound-pressure level. That is, a 20W amplifier driving the 94dB-sensitive speaker will produce the same sound-pressure level as 160W driving the 85dB-sensitive speaker.

This knowledge allows us to match the amplifier's output power to the loudspeaker's sensitivity and get the best possible performance for the money. In a system employing an amplifier with not enough output power for a particular loudspeaker, the sound will lack dynamics, distort on musical peaks, have soft and sluggish bass, and sound constricted. If the amplifier has an excess of power for a particular loudspeaker, the penalty is not sonic but financial; you are paying for watts in the amplifier you'll never use. An amplifier's cost is often proportional to its output power—watts equal dollars. And wasted watts represent money that could have been better spent on other components in your system.

Before we look at some real-world examples of matching amplifier power to a loudspeaker's sensitivity, there are a few more factors to consider.

The first is room size. The bigger the room, the more amplifier power you'll need. A rough guide suggests that quadrupling the room volume requires a doubling of amplifier power to achieve the same sound-pressure level. How acoustically reflective or absorptive your listening room is will also affect the best size of amplifier for your system. If we put the same-sensitivity loudspeakers in two rooms of the same size, one room acoustically dead (absorptive) and the other acoustically live (reflective), we would need roughly double the amplifier power to achieve the same sound-pressure level in the dead room as in the live room.

Finally, how loudly you listen to music greatly affects how much amplifier power you need. Chamber music played softly requires much less amplifier power than rock or orchestral music played loudly.

We can see that a low-sensitivity loudspeaker, driven by orchestral music in the large, acoustically dead room of someone who likes high playback levels, may require nearly one-hundred times the amplifier power needed by someone listening to chamber music at moderate listening levels through high-sensitivity loudspeakers in a small, live room. A 10Wpc amplifier may satisfy the second listener; the first listener may need 750Wpc.

Now for some real-world examples. A loudspeaker of 88dB sensitivity in an average-sized room (say, 4000 cubic feet) featuring acoustics typical of today's homes requires a minimum of about 50Wpc. If you enjoy orchestral music, which has a very wide dynamic range, an amplifier with a rating of 75Wpc would be more appropriate.

Choosing an appropriate amplifier power-output range for your loudspeakers, listening tastes, room, and budget is essential to getting the best sound for your money. If the amplifier is under-powered for your needs, you'll never hear the system at its full potential. The sound will be constricted, fatiguing, lack dynamics, and the music will have a sense of strain on climaxes. Conversely, if you spend too much of your budget on a bigger amplifier than you need, you may be shortchanging other components. Choosing just the right amplifier power is of paramount importance.

The sure-fire way of determining if a particular amplifier works well with a particular loudspeaker is to audition the combination, either in your home or the dealer's showroom. Later in this chapter ("What to Listen For") we'll learn how to determine by listening whether an amplifier is up to the task of driving a particular loudspeaker.

Output-Power Specifications: Read the Fine Print

Manufacturers use all sorts of tricks to make their amplifiers seem as powerful as possible on the specification sheet. When comparing amplifier power ratings, make sure the specified power is continuous or RMS rather than peak. Some manufacturers will claim a power output of 200W, for example, but not specify whether that power output is available only during transient musical events such as drum beats, or if the amplifier can deliver that power continuously into a load. Also be sure the power rating is specified into an 8-ohm impedance. Rating the amplifier's power into 4 ohms makes the amplifier seem more powerful than it really is.

Why Two "100W" Amplifiers Can Have Different Output Powers

It is possible to buy an audio/video receiver today with a rated power of 100Wpc into 8 ohms across all five of its amplifier channels for less than $300. You can also spend $4000 for a stereo amplifier that is identically rated at 100Wpc into 8 ohms. Do these amplifiers have the same output power?

Yes and no. If both amplifiers are driving a continuous signal such as a test tone into 8 ohms, as occurs on a test bench, they both deliver 100W. But music is not a continuous signal, and loudspeakers rarely have an impedance of 8 ohms at all frequencies. Most loudspeakers have dips in their impedance; a speaker rated at 8 ohms could present a 4-ohm impedance to the amplifier over most of the audio-frequency band. Some speakers even have impedance dips as low as 2 ohms. All solid-state amplifiers can increase their output powers into a 4-ohm speaker relative to an 8-ohm speaker, but they vary in the amount of power increase. For example, the audio/video receiver may be able to deliver perhaps 140W into 4 ohms, but the high-end stereo amplifier can deliver 200W into 4 ohms. This difference has real-world consequences. The ability to increase output power into low impedances indicates how much current the amplifier can deliver to the loudspeaker. It is current flow through the loudspeakers' voice coils (in dynamic loudspeakers) that creates the electromagnetic force that causes the cones and domes to move, producing sound. If current flow through the voice coil is constrained, so is the music.

In addition, higher-quality amplifiers have greater *dynamic headroom*. This term describes an amplifier's ability to deliver instantaneous bursts of short-term power well above the amplifier's rated continuous power. For example, snare drum beats that require 180W for a few milliseconds would be reproduced cleanly by a 100-watt amplifier with good dynamic headroom, but sound compressed, lifeless, and even distorted by a lesser-quality 100-watt amplifier with very little dynamic headroom, even if the amplifiers' continuous power ratings were identical.

The result is that two "100W" amplifiers can have very different power-delivery characteristics—and musical performances—when driving loudspeakers under real-world conditions.

A hallmark of a high-end amplifier is the ability to deliver performance beyond the specification sheet. The high-end amplifier is designed to reproduce music as faithfully as possible under real-world conditions, not to look good "on paper." An amplifier's ability to increase its output power into a low impedance, and deliver good dynamic headroom, is related to two aspects of its design: the *power supply* and *output stage.* The power supply is the circuit in the amplifier that converts the AC power from your wall into direct current that feeds the amplifier's circuits. Most of a high-end amplifier's weight is in the power supply. The amplifier's output stage is a bank of transistors (or tubes) that do the actual work of pushing and pulling electrical current through the voice coils of your loudspeakers. To achieve the exemplary performance I've described, the power supply needs to be large and heavy, and the output stage must employ multiple heavy-duty transistors cooled by large heat sinks (the cooling fins visible on many amplifiers). Power supplies, output transistors, and heat sinks are by far the most expensive components in a power amplifier; skimping on these saves the manufacturer money at the expense of real-world musical performance. High-end amplifiers are easily distinguishable from mid-fi amplifiers of comparable power rating by their large power supplies and ample heat sinks. These are just two aspects of why high-end designs sound better than mass-market products, but there are countless other design elements throughout the high-end amplifier that make it a more faithful device for reproducing music.

Other Power-Amplifier Considerations

Balanced Inputs

If your preamp has balanced outputs, you may want to consider a power amplifier with balanced inputs. Most power amps with balanced inputs also provide unbalanced inputs, allowing you to compare these two connection methods before deciding which one to leave in your system. Some preamplifier/power-amplifier combinations sound better via balanced connection, others via the unbalanced jacks. The best way to discover which method is better is by listening to both. (See Chapter 7 for a discussion of why some preamplifiers may sound better from their unbalanced outputs.) Borrow a pair of interconnects from your dealer so you can make this comparison.

Bridging

Some stereo power amplifiers can be "bridged" to function as monoblocks. Bridging configures a stereo amplifier to function as a more powerful single-channel amplifier. The amplifier will have a switch on the rear panel to convert the amplifier to bridged operation. Note that two bridged amplifiers are needed for stereo. If you have a stereo amplifier that can be bridged and you want more power, simply buy a second, identical amplifier and bridge the two for more total power.

Bridging changes the amplifier's internal connections so that one channel amplifies the positive half of the waveform and the other channel amplifies the negative half. The loudspeaker is connected as the "bridge" between the two amplifier channels instead of between one channel's output and ground.

What to Listen For

How can you tell if the power amplifier you're considering will work well with your loudspeakers? Simple: Borrow the amplifier from your dealer for the weekend and listen to it. This is the best way of not only assessing its musical qualities, but determining how well it drives your loudspeakers. In addition, listening to the power amplifier at home will let you hear if the product's sonic signature complements the rest of your system.

The next best choice is if the dealer sells the same loudspeakers you own and allows you to audition the combination in the store. If neither of these options is practical, consider bringing your loudspeakers into the store for a final audition.

All the sonic and musical characteristics described in Chapter 4 apply to power amplifiers. However, some sonic characteristics are more influenced by the power amplifier than by other components.

The first thing to listen for is whether the amplifier is driving the loudspeakers adequately. The most obvious indicator is bass performance. If the bass is soggy and slow, or lacks punch, the amplifier probably isn't up to the job of driving your loudspeakers. Weak bass is a sure indication that the amplifier is underpowered for a particular pair of speakers because the woofer puts the greatest current demands on the amplifier. Other telltale signs that the amplifier is running out of power include loss of dynamics, a sense of strain on musical peaks, hardening of timbre, reduced sense of pace and rhythm, and soundstage collapse or congestion. Let's look at each of these individually.

First, play the system at a moderate volume. Select music with a wide dynamic range—either full orchestral music with a loud climax accompanied by bass drum, or music with a powerful rhythmic drive from bass guitar and kickdrum working together. Audiophile recordings typically have much wider dynamic ranges than general-release discs, making them a better source for evaluation. Music that has been highly compressed to play over the radio on a 3" car speaker will tell you less about what the system is doing dynamically.

After you've become used to the sound at a moderate level, increase the volume—you want to push the amplifier to find its limits. Does the bass seem to give out when you turn it up, or does the amplifier keep on delivering? Listen to the dynamic impact of kickdrum on a recording with lots of bottom-end punch. It should maintain its tightness, punch, quickness, and depth at high volume. If it starts to sound soggy, slow, or loses its power, you've gone beyond the amplifier's comfortable operating point. After a while, you can get a feel for when the amplifier gets into trouble. Is the sound strained on peaks, or effortless?

NOTE: When performing this experiment, be sure not to overdrive your loudspeakers. Turn down the volume at the first sign of loudspeaker overload (distortion or a popping sound).

Chapter 8

Compare the amplifier's sound at high and low volumes. Listen for brass instruments becoming hard and edgy at high volumes. See if the soundstage degenerates into a confused mess during climaxes. Does the bass drum lose its power and impact? An excellent power amplifier operating near its maximum power capability will preserve the senses of space, depth, and focus, while maintaining liquid instrumental timbre. Moreover, adequate power will produce a sense of ease; lack of power often creates listener fatigue. Music is much more enjoyable through a power amplifier with plenty of reserve power.

All the problems I've just described are largely the result of the amplifier running out of current. Just where this happens is a function of the amplifier's output power, its ability to deliver current into the loudspeaker, the loudspeaker's sensitivity and impedance, the room size, and how loudly you expect your hi-fi system to play. Even when not pushed to its maximum output, a more powerful amplifier will often have a greater sense of ease, grace, and dynamics than a less powerful amplifier.

9

Loudspeakers

Of all the components in your audio system, the loudspeaker's job is by far the most difficult. The loudspeaker is expected to reproduce the sound of a pipe organ, the human voice, and a violin through the same electromechanical device—all at the same levels of believability, and all at the same time. The tonal range and complex harmonic structure of virtually every instrument in the orchestra is to be reproduced from a relatively tiny box. Loudspeakers that are part of a home-theater system are also expected to reproduce car crashes, tornadoes, and explosions—sometimes simultaneously with dialogue.

It's no wonder that loudspeaker designers spend their lives battling the laws of physics to produce musical and practical loudspeakers. Unlike other high-end designers who create a variety of products, the loudspeaker designer is singular in focus, dedicated in intent, and deeply committed to the unique blend of science and art that is loudspeaker design.

Although even the best loudspeakers can't convince us that we're hearing live music, they nonetheless are miraculous in what they *can* do. Think about this: A pair of loudspeakers converts two two-dimensional electrical signals into a three-dimensional "soundspace" spread out before the listener. Instruments seem to exist as *objects* in space; we hear the violin here, the brass over there, and the percussion behind the other instruments. A vocalist appears as a palpable, tangible image exactly between the two loudspeakers. The front of the listening room seems to disappear, replaced by the music. It's so easy to close your eyes and be transported into the musical event.

To achieve this experience in your home, however, you must carefully choose the best loudspeakers from among the literally thousands of models on the market. As we'll see, choosing loudspeakers is a challenging job.

How to Choose a Loudspeaker

The world abounds in poor-quality, even dreadful, loudspeakers. What's more, some very bad loudspeakers are expensive, while superlative models may sell for a fraction of an inferior model's price. There is sometimes little relationship between price and musical performance.

This situation offers the loudspeaker shopper both promise and peril. The promise is of finding an excellent loudspeaker for a reasonable price. The peril is of sorting through mediocre models to find the rare gems that offer either high absolute performance, or sound quality far above what their price would indicate.

This is where reviews come in handy. Reviewers who write for audio magazines hear lots of loudspeakers (at dealers, trade shows, and consumer shows), but review only those that sound promising. This weeds out the vast majority of underachievers. Of the loudspeakers that *are* reviewed, some are found to be unacceptably flawed, others are good for the money, while a select few are star overachievers that clearly outperform their similarly priced rivals.

The place to start loudspeaker shopping, therefore, is in the pages of a reputable magazine with high standards for what constitutes good loudspeaker performance. Be wary of magazines that end every review with a "competent for the money" recommendation. Not all loudspeakers are good; therefore, not all reviews should be positive. The tone of the reviews—positive or negative—should reflect the wide variation in performance and value found in the marketplace.

After you've read lots of loudspeaker reviews, make up your short list of products to audition from the *crème de la crème*. There are several criteria to apply in making this short list to ensure that you get the best loudspeaker for your individual needs. As you apply each criterion described, the list of candidate loudspeakers will get shorter and shorter, thus easing your decision-making process. If you find yourself with too few choices at the end of the process, go back and revise your criteria. For example, if you find a loudspeaker that's perfect in all ways but size, you may want to find the extra space in your living room. Similarly, an ideal loudspeaker costing a little more than you planned to spend may suggest a budget revision. As you go through this selection process, remember that the perfect loudspeaker for you is probably out there. Be selective and have high standards. You'll be rewarded by a much higher level of musical performance than you thought you could afford.

Your first decision in choosing loudspeakers is whether your system will reproduce 2-channel music, multichannel music, film soundtracks as part of a home-theater system, or all of these. No matter what loudspeaker configuration you choose, the three starting criteria for making your short list of candidates will be the same. Let's look at those criteria:

1) Size, Appearance, and Integration in the Home

After you've designated a place for your loudspeakers, determine the optimum loudspeaker size for your room—the urban apartment dweller will likely have tighter size constraints than the suburban audiophile. Some listeners will want the loudspeakers to discreetly blend into the room; others will make the hi-fi system the room's center of activity and won't mind large, imposing loudspeakers. When choosing a place for your loudspeakers, keep in mind that their placement is a crucial factor in how your system will sound. (Chapter 12 includes an in-depth treatment of loudspeaker positioning.)

The loudspeaker's appearance is also a factor to consider. An inexpensive, vinyl-covered box would be out of place in an elegantly furnished home. Many high-end loudspeakers are finished in beautiful cabinetry that will complement any decor. This level of finish can, however, add greatly to the loudspeaker's price.

If you don't have room for full-range, floorstanding speakers, consider a separate subwoofer/satellite system. This is a loudspeaker system that puts the bass-reproducing driver in an enclosure you can put nearly anywhere, and the midrange- and treble-reproducing elements in a small, unobtrusive cabinet. You'll still get a full sound, but without the visual domination of your living room that often goes with floorstanding speakers. Moreover, the satellite speakers' small cabinets often help them achieve great soundstaging. An added benefit is that you can position the woofer cabinet for best bass performance and the satellites for best soundstaging.

If your room won't easily accommodate floorstanding or stand-mounted loudspeakers, consider on-wall or in-wall models. As its name suggests, the in-wall is mounted inside a cavity cut into drywall (Fig.9-1). The on-wall is generally a very shallow, slender speaker that mounts to the wall (Fig.9-2). The popularity of flat-panel television (LCD and plasma) has fueled the growth of on-wall models that complement a flat-panel video display. Although in-wall and on-wall loudspeakers have improved dramatically in the past few years, they will not deliver the same sound quality as a good free-standing loudspeaker.

Fig.9-1 In-wall loudspeakers can be mounted unobtrusively. (Courtesy Bowers & Wilkins Loudspeakers)

Chapter 9

Fig.9-2 On-wall loudspeakers create a sleek, low-profile look. (Courtesy Bowers & Wilkins Loudspeakers)

2) Match the Loudspeaker to Your Electronics

The loudspeaker should be matched to the rest of your system, both electrically and musically. A loudspeaker that may work well in one system may not be ideal for another system—or another listener.

Let's start with the loudspeaker's electrical characteristics. The power amplifier and loudspeaker should be thought of as an interactive combination; the power amplifier will behave differently when driving different loudspeakers. Consequently, the loudspeaker should be chosen for the amplifier that will drive it.

The first electrical consideration is a loudspeaker's sensitivity—how much sound it will produce for a given amount of amplifier power. Loudspeakers are rated for sensitivity by measuring their sound-pressure level (SPL) from one meter away while they are being fed one watt (1W) of power. For example, a sensitivity specification of "88dB, 1W/1m" indicates that this particular loudspeaker will produce a sound-pressure level of 88dB when driven with an input power of 1W, measured at a distance of 1m.

As we saw in the previous chapter, a loudspeaker's sensitivity is a significant factor in determining how well it will work with a given power-amplifier output wattage. To produce a loud sound (100dB), a loudspeaker rated at 80dB sensitivity would require 100W. A loudspeaker with a sensitivity of 95dB would require only 3W to produce the same sound-pressure level. Each 3dB decrease in sensitivity requires double the amplifier power to produce the same SPL. (This is discussed in greater technical detail in Chapter 8, "Power and Integrated Amplifiers.")

Another electrical factor to consider is the loudspeaker's *load impedance*. This is the electrical resistance the power amp meets when driving the loudspeaker. The lower the loudspeaker's impedance, the more demand is placed on the power amp. If you choose low-impedance loudspeakers, be certain the power amp will drive them adequately. Keep in mind that a speaker rated at "8 ohms" will probably not present an 8-ohm load to the amplifier at all frequencies. Nearly all loudspeakers have dips in the impedance at certain frequencies, making them harder to drive.

On a musical level, you should select a loudspeaker as sonically neutral as possible. If you have a bright-sounding CD player or power amp, it's a mistake to buy a loudspeaker that sounds soft or dull in the treble to compensate. Instead, change your CD player or amplifier.

Another mistake is to drive high-quality loudspeakers with poor amplification or source components. The high-quality loudspeakers will resolve much more information than lesser loudspeakers—including imperfections in the electronics and source components. All too many audiophiles drive great loudspeakers with mediocre source components and never realize their loudspeakers' potential. Match the loudspeakers' quality to that of the rest of your system. (Use the guidelines in Chapter 3 to set a loudspeaker budget within the context of the cost of your entire system.)

3) Musical Preferences and Listening Habits

If the perfect loudspeaker existed, it would work equally well for chamber music and heavy metal. But because the perfect loudspeaker remains a mythical beast, musical preferences must play a part in choosing a loudspeaker. If you listen mostly to small-scale classical music, choral works, or classical guitar, a minimonitor would probably be your best choice. Conversely, rock listeners need the dynamics, low-frequency extension, and bass power of a large full-range system. Different loudspeakers have strengths and weaknesses in different areas; by matching the loudspeaker to your listening tastes, you'll get the best performance in the areas that matter most to you.

Other Guidelines in Choosing Loudspeakers

In addition to these specific recommendations, there are some general guidelines you should follow in order to get the most loudspeaker for your money.

First, buy from a specialty audio retailer who can properly demonstrate the loudspeaker, advise you on system matching, and tell you the pros and cons of each candidate. Many high-end audio dealers will let you try the loudspeaker in your home with your own electronics and music before you buy.

Take advantage of the dealer's knowledge—and reward him with the sale. It's not only unfair to the dealer to use his or her expensive showroom and knowledgeable salespeople to find out which product to buy, then look for the loudspeaker elsewhere at a lower price; it also prevents you from establishing a mutually beneficial relationship with him or her.

In general, loudspeakers made by companies that make *only* loudspeakers are better than those from companies who also make a full line of electronics. Loudspeaker design may be an afterthought to the electronics manufacturer—something to fill out the line. Conversely, many high-end loudspeaker companies have an almost obsessive dedication to the art of loudspeaker design. Their products' superior performance often reflects this commitment.

Most of the hundreds of speaker manufacturers buy raw drivers from just a few companies and assemble those drivers into their own cabinets. This makes it relatively easy to start a loudspeaker company. Indeed, the speaker industry is full of small start-ups that create a new design in the hope of joining the ranks of established companies.

At the next level are companies who make high-end loudspeakers in relatively small quantity, but have been in business for some time and have an established dealer base.

Many companies who started at this level have been graduated to the establishment of the high-end. They have moved to relatively large-scale manufacturing yet still maintain high quality. These companies are on the threshold between craft-based manufacturing and full-scale industrial manufacturing.

There's also a class of speaker manufacturer that grew out of its high-end roots to become a mass-market manufacturer. Some of these companies long ago dropped any pretense of being high-end and have instead focused on cutting costs and appealing to the widest possible audience. But some speaker companies founded on high-end principles have retained their dedication to quality despite becoming mass-manufacturers. These companies combine thoughtful and caring design with economy-of-scale manufacturing to deliver products that offer very high performance and extraordinary value. Such companies might not produce state-of-the-art loudspeakers, or even those considered at the upper end of what's possible in music reproduction, but they consistently deliver superlative sound, solid build quality, and nice cabinetry at reasonable prices. These companies are still driven by a passion for great sound. Such companies are worth your interest.

Don't buy a loudspeaker based on technical claims. Some products claiming superiority in one aspect of their performance may overlook other, more important aspects. Loudspeaker design requires a balanced approach, not reliance on some new "wonder" technology that may have been invented by the loudspeaker manufacturer's marketing department. Forget about the technical hype and listen to how the loudspeaker reproduces *music*. You'll hear whether or not the loudspeaker is any good.

Don't base your loudspeaker purchases on brand loyalty or longevity. Many well-known and respected names in loudspeaker design of 20 years ago are no longer competitive. Such a company may still produce loudspeakers, but its recent products' inferior performance only throws into relief the extent of the manufacturer's decline. The brands the general public thinks represent the state of the art are actually among the worst-sounding loudspeakers available. These companies were either bought by multinational business conglomerates that didn't care about quality and just wanted to exploit the brand name, or the company has forsaken high performance for mass-market sales. Ironically, the companies with the most visible advertisements in the mass media often offer the lowest-performance loudspeakers.

The general public also believes that the larger the loudspeaker and the more drivers it has, the better it is. Given the same retail price, there is often an *inverse* relationship between size/driver count and sonic performance. A good two-way loud-speaker—one that splits the frequency spectrum into two parts for reproduction by a woofer and a tweeter—with a 6" woofer/midrange and a tweeter in a small cabinet is likely to be vastly better than a similarly priced four-way in a large, floorstanding enclosure. Two high-quality drivers are much better than four mediocre ones. Further, the larger the cabinet, the more difficult and expensive it is to make it free from vibrations that degrade the sound. The four-way speaker's more extensive crossover will require more parts; the two-way can use just a few higher-quality crossover parts. The large loudspeaker will probably be unlistenable; the small two-way may be superbly musical.

If both of these loudspeakers were shown in a catalog and offered at the same price, however, the large, inferior system would outsell the high-quality two-way by at least 10 times. The perceived value of more hardware for the same money is much higher.

The bottom line: You can't tell anything about a loudspeaker until you listen to it. In the next section, we'll examine common problems in loudspeakers and how to choose one that provides the highest level of musical performance.

Finding the Right Loudspeaker—Before You Buy

You've done your homework, read reviews, and narrowed down your list of candidate loudspeakers based on the criteria described earlier—you know what you want. Now it's time to go out and listen. This is a crucial part of shopping for a loudspeaker, and one that should be approached carefully. Rather than buying a pair of speakers on your first visit to a dealer, consider this initial audition to be simply the next step. Don't be in a hurry to buy the first loudspeaker you like. Even if it sounds very good to you, you won't know how good it is until you've auditioned several products.

Audition the loudspeaker with a wide range of familiar recordings *of your own choosing*. Remember that a dealer's strategic selection of music can highlight a loud-speaker's best qualities and conceal its weaknesses—after all, his job is to present his products in the best light. Further, auditioning with only audiophile-quality recordings won't tell you much about how the loudspeaker will perform with the music you'll be playing at home, most of which was likely *not* recorded to high audiophile standards. Still, audiophile recordings are excellent for discovering specific performance aspects of a loudspeaker. The music selected for auditioning should therefore be a combination of your favorite music, and diagnostic recordings chosen to reveal different aspects of the loudspeaker's performance. When listening to your favorite music, forget about specific sonic characteristics and pay attention to how much you're enjoying the sound. Shift into the analytical mode only when playing the diagnostic recordings. Characterize the sound quality according to the sonic criteria described in Chapter 4, and later in this chapter.

Visit the dealer when business is slow so you can spend at least an hour with the loudspeaker. Some loudspeakers are appealing at first, and then lose their luster as their flaws begin to emerge over time. The time to lose patience with the speakers is in the dealer's showroom, not a week after you've bought them. And don't try to audition more than two sets of loudspeakers in a single dealer visit. If you must choose between three models, select between the first two on one visit, then return to compare the winner of the first audition with the third contender. You should listen to each candidate as long as you want (within reason) to be sure you're making the right purchasing decision.

Make sure the loudspeakers are driven by electronics and source components of comparable quality to your components. It's easy to become infatuated with a delicious sound in a dealer's showroom, only to be disappointed when you connect the loudspeakers to lesser quality electronics. Ideally, you should drive the loudspeakers under audition with the same level of power amp as you have at home, or as you intend to buy with the loudspeakers.

Of course, the best way to audition loudspeakers is in your own home—you're under no pressure, you can listen for as long as you like, and you can hear how the loudspeaker performs with your electronics and in your listening room. Home audition removes much of the guesswork from choosing a loudspeaker. But because it's impractical to take every contender home, and because many dealers will not allow this, save your home auditioning for only those loudspeakers you are seriously considering.

If you are auditioning a multichannel speaker system and the dealer is demonstrating the system with both movies and music, give greater weight to your opinion of the system's performance with music. It's fairly easy to impress a listener with the artificial sounds of a movie soundtrack—special effects, explosions, and car crashes. It is much more difficult for a loudspeaker system to reproduce natural sounds such as the human voice, a violin, an acoustic guitar, and other instruments. Moreover, identifying loudspeaker colorations (departures from accuracy) is much easier with recordings of real instruments.

What to Listen For

There are several common flaws in loudspeaker performance that you should listen for. Though some of these flaws are unavoidable in the lower price ranges, a loudspeaker exhibiting too many of them should be quickly passed over.

Listen for thick, slow, and tubby bass. One of the most annoying characteristics of poor loudspeakers is colored, peaky, and pitchless bass. You should hear distinct pitches in bass notes, not a low-frequency, "one-note" growling under the music. Male speaking voice is a good test for upper-bass colorations; it shouldn't have an excessive or unnatural chesty sound.

Certain bass notes shouldn't sound louder than others. Listen to solo piano with descending or ascending lines played evenly in the instrument's left-hand, or lower, registers. Each note should be even in tone and volume, and clearly articulated. If one note sounds different from the others, it's an indication that the loudspeaker may have a problem at that frequency.

The bottom end should be tight, clean, and "quick." When it comes to bass, quality is more important than quantity. Poor-quality bass is a constant reminder that the music is being artificially reproduced, making it that much harder to hear only the music and not the loudspeakers. The paradigm of what bass should *not* sound like is a "boom truck." Those car stereos are designed for maximum output at a single frequency, not articulate and tuneful bass. Unfortunately, *more* bass is generally an indicator of *worse* bass performance in low- to moderately-priced loudspeakers. A lean, tight, and articulate bass is preferable in the long run to the plodding boominess that characterizes inferior loudspeakers.

Listen to kickdrum and bass guitar working together. You should hear the bass drum's dynamic envelope through the bass guitar. The drum should lock in rhythmically rather than seem to lag slightly behind the bass guitar. A loudspeaker that gets this wrong dilutes rhythmic power, making the rhythm sound sluggish, even slower. But when you listen to a loudspeaker that gets this right, you'll find your foot tapping and hear a more "upbeat" and involving quality to the music.

Midrange coloration is a particularly annoying problem with some loudspeakers. Fortunately, coloration levels are vastly lower in today's loudspeakers than they were even 15 years ago. Still, there are lots of colored loudspeakers out there. These can be identified by their "cupped hands" coloration on vocals, a nasal quality, or an emphasis on certain vowel sounds. A problem a little higher in frequency is manifested as a "clangy" piano sound. A good loudspeaker will present vocals as pure, open, and seeming to exist independently of the loudspeakers. Midrange problems will also make the music sound as though it is coming out of boxes rather than existing in space.

Poor treble performance is characterized by grainy or dirty sound to violins, cymbals, and vocal sibilants (*s* and *sh* sounds). Cymbals should not splash across the soundstage, sounding like bursts of noise. Instead, the treble should be integrated with the rest of the music and not call attention to itself. The treble shouldn't sound hard and metallic; instead, cymbals should have some delicacy, texture, and pitch. If you find that a pair of speakers is making you aware of the treble as a separate component of the music, keep looking.

Another thing to listen for in loudspeakers is their ability to play loudly without congestion. The sounds of some loudspeakers will be fine at low levels, but will congeal and produce a giant roar when pushed to high volumes. Listen to orchestral music with crescendos—the sound should not collapse and coarsen during loud, complex passages.

Finally, the loudspeakers should "disappear" into the soundstage. A good pair of loudspeakers will unfold the music in space before you, giving no clue that the sound is coming from two boxes placed at opposite sides of the room. Singers should be heard as pinpoint, palpable images directly between the loudspeakers (if that's how they've been recorded). The sonic image of an instrument should not "pull" to one side or another when the instrument moves between registers. The music should sound open and transparent, not thick, murky, or opaque. Overall, the less you're aware of the loudspeakers themselves, the better.

Chapter 9

When the speaker is playing loudly, put your hand on the cabinet and feel if it is vibrating. The less vibration, the better. The best loudspeakers will exhibit no apparent cabinet vibration whatsoever.

When auditioning a multichannel loudspeaker package, pay special attention to the center speaker's performance. The center speaker plays an important role in home theater; specifically, it reproduces nearly all of the film's dialogue. Listen for how clearly you can hear the actors speaking (called dialogue intelligibility), particularly when the musical score or special effects compete with the dialogue. Select a scene on a DVD that is particularly challenging and play this scene over the loudspeakers you are considering.

Some loudspeakers with less-than-high-end aspirations have colorations intentionally designed into them. The bass is made to be big for "warmth," the treble excessively bright to give the illusion of "clarity." Such speakers are usually extremely sensitive, so that they'll play loudly in comparisons made without level matching. These loudspeakers may impress the unwary in a two-minute demonstration, but will become extremely annoying not long after you've brought them home. You're unlikely to find such products in a true high-end audio store.

Finally, the surest sign that a loudspeaker will provide long-term musical satisfaction at home is if, during the audition, you find yourself greatly enjoying the music and not thinking about loudspeakers at all.

The flaws described here are only the most obvious loudspeaker problems; a full description of what to listen for in reproduced music in general is found in Chapter 4.

Loudspeakers for Home Theater

Although we'll cover the role of each loudspeaker in a multichannel audio system in Chapter 10, a brief overview here will be helpful.

Film soundtracks are mixed into 5.1 audio channels and carried to your home by the Dolby Digital and DTS surround-sound formats. The 5.1 channels are reproduced by five loudspeakers in your living room, plus an optional subwoofer (the ".1" channel). The five loudspeakers are *left, center, right, surround left,* and *surround right*. The left and right speakers reproduce most of the film soundtrack's music and special effects. The center speaker carries the film's dialogue and some special effects. It also anchors on-screen sounds on the screen. The left, right, and center speakers are together called the front speakers, because they are positioned at the front of your room, with the left and right speakers on either side of your video display and the center speaker mounted on top of, or just beneath, the video display. The surround speakers are mounted on the side or rear walls behind the listening position. Their job is to reproduce atmospheric sounds and some special effects. Because surround speakers receive much less signal (and thus amplifier power) than the front speakers, they can be small and unobtrusive.

The subwoofer is optional if you use full-range floorstanding speakers, but a requirement if your five main speakers are small satellites. The subwoofer is usually driven by the ".1" signal in the film soundtrack, a channel reserved for high-impact, low-bass sounds. The subwoofer signal also includes the bass from any channel in which you have designated SMALL in the receiver or controller's set-up menu. This prevents low-bass from overloading small speakers.

Loudspeaker Types and How They Work

Many mechanisms for making air move in response to an electrical signal have been tried over the years. Three methods of creating sound work well enough—and are practical enough—to be used in commercially available products. These are the *dynamic driver*, the *ribbon transducer*, and the *electrostatic panel*. A loudspeaker using dynamic drivers is often called a *box* loudspeaker because the drivers are mounted in a box-like enclosure or cabinet. Ribbon and electrostatic loudspeakers are called *planar* loudspeakers because they're usually mounted in flat, open panels. The term transducer describes any device that converts energy from one form to another. A loudspeaker is a transducer because it converts electrical energy into sound.

The Dynamic Driver

The most popular loudspeaker technology is without question the dynamic driver. Loudspeakers using dynamic drivers are identifiable by their familiar cones and domes. The dynamic driver's popularity is due to its many advantages: wide dynamic range, high power handling, high sensitivity, relatively simple design, and ruggedness. Dynamic drivers are also called *point-source* transducers because the sound is produced from a point in space.

Dynamic loudspeaker systems use a combination of different-sized dynamic drivers. The low frequencies are reproduced by a paper or plastic cone *woofer*. High frequencies are generated by a *tweeter*, usually employing a small metal or fabric dome. Some loudspeakers use a third dynamic driver, the *midrange*, to reproduce frequencies in the middle of the audio band.

Despite the very different designs of these drivers, they all operate on the same principle (Fig.9-3).

Electrical current from the power amplifier flows through the driver's *voice coil*. This current flow sets up a magnetic field around the voice coil that expands and collapses at the same frequency as the audio signal. The voice coil is suspended in a permanent magnetic field generated by magnets in the driver. This permanent magnetic field interacts with the magnetic field generated by current flow through the voice coil, alternately pushing and pulling the voice coil back and forth. Because the voice coil is attached to the driver's cone, this magnetic interaction pulls the cone back and forth, producing sound. Dynamic drivers are also called *moving-coil* drivers, for obvious reasons.

Other elements of the dynamic driver include a *spider* that suspends the voice coil in place as it moves back and forth. The *basket* is a cast- or stamped-metal structure holding the entire assembly together. (Cast baskets are generally found in higher-quality loudspeakers, stamped baskets in budget models.) A ring of compliant rubber material, called the *surround*, attaches the cone to the basket rim. The surround allows the cone to move back and forth while still attached to the basket. The maximum distance the cone moves back and forth is called its *excursion*.

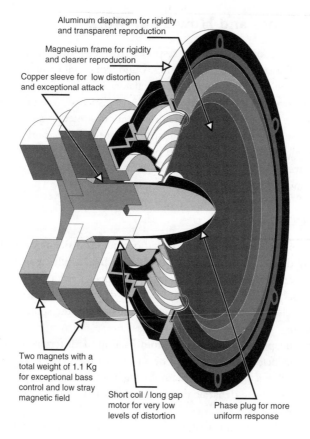

Aluminum diaphragm for rigidity and transparent reproduction

Magnesium frame for rigidity and clearer reproduction

Copper sleeve for low distortion and exceptional attack

Two magnets with a total weight of 1.1 Kg for exceptional bass control and low stray magnetic field

Short coil / long gap motor for very low levels of distortion

Phase plug for more uniform response

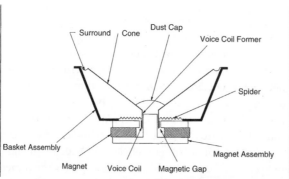

Surround   Cone   Dust Cap

Voice Coil Former

Spider

Basket Assembly

Magnet Assembly

Magnet   Voice Coil   Magnetic Gap

Fig.9-3 Dynamic driver cut-away and cross section. (Courtesy Thiel Audio)

Common cone materials include paper, paper impregnated with a stiffening agent, a form of plastic such as polypropylene, or exotic materials such as Kevlar. Metal (including titanium) has also been used in woofer cones, as have sandwiched layers of different materials. Designers use these materials to prevent a form of distortion called *breakup*. Breakup occurs when the cone material flexes instead of moving as a perfect piston.

Tweeters work on the same principle, but typically use a 1" dome instead of a large cone. Common dome materials include plastic, woven fiber coated with a rubbery material, titanium, aluminum and aluminum alloys, and gold-plated aluminum. Diamond tweeter diaphragms have also been used in extremely expensive loudspeakers. Unlike cone drivers, which are driven at the cone's apex, dome diaphragms are driven at the dome's outer perimeter.

Midrange drivers are smaller versions of the cone woofer. Some, however, use dome diaphragms instead of cones.

The Planar-Magnetic Transducer

The next popular driver technology is the planar-magnetic transducer, also known as a ribbon driver. Although the terms "ribbon" and "planar magnetic" are often used interchangeably, a true ribbon driver is actually a sub-class of the planar-magnetic driver. Let's look at a true ribbon first.

Instead of using a cone attached to a voice coil suspended in a magnetic field, a ribbon driver uses a strip of material (usually aluminum) as a diaphragm suspended between the north-south poles of two magnets (Fig.9-4). The ribbon is often pleated for additional strength. The audio signal travels through the electrically conductive ribbon, creating a magnetic field around the ribbon that interacts with the permanent magnetic field. This causes the ribbon to move back and forth, creating sound. In effect, the ribbon functions as both the voice coil and the diaphragm. The ribbon can be thought of as the voice coil stretched out over the ribbon's length.

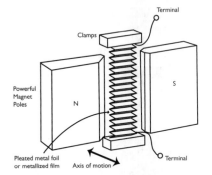
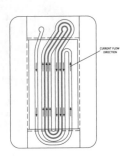

Fig.9-4 In a true ribbon driver, the audio signal flows through the ribbon (left). In a planar-magnetic driver, a conductor is bonded to the back of a diaphragm (right). (Right diagram courtesy BG Corp.)

In all other planar-magnetic transducers, a flat or slightly curved diaphragm is driven by an electromagnetic conductor. This conductor, which is bonded to the back of the diaphragm, is analogous to a dynamic driver's voice coil, here stretched out in straight-line segments. In most designs, the diaphragm is a sheet of plastic with the electrical conductors bonded to its surface. The flat metal conductor provides the driving force, but occupies only a portion of the diaphragm area. Such drivers are also called *quasi-ribbon* transducers (Fig.9-4, right diagram, on the previous page).

A planar driver is a true ribbon only if the diaphragm is conductive and the audio signal flows directly through the diaphragm, rather than through conductors bonded to a diaphragm, as in quasi-ribbon drivers. (Despite this semantic distinction, I'll use the term "ribbon" throughout the rest of this section, with the understanding that it covers both true ribbons and quasi-ribbon drivers.)

Ribbon drivers like the one in Fig. 9-4 (left) are called *line-source* transducers because they produce sound over a line rather than from a point, as does a dynamic loudspeaker.

The main technical advantage of a ribbon over a moving-coil driver is the ribbon's vastly lower mass. Instead of using a heavy cone, voice coil, and voice-coil former to move air, the only thing moving in a ribbon is a very thin strip of aluminum. A ribbon tweeter may have one quarter the mass and 10 times the radiating area of a dome tweeter's diaphragm. Low mass is a high design goal: the diaphragm can respond more quickly to transient signals. In addition, a low-mass diaphragm will stop moving immediately after the input signal has ceased. The ribbon starts and stops faster than a dynamic driver, allowing it to more faithfully reproduce transient musical information.

The ribbon driver is usually mounted in a flat, open-air panel that radiates sound to the rear as well as to the front. A loudspeaker that radiates sound to the front and rear is called a *dipole,* which means "two poles." Fig.9-5 shows the radiation patterns of a point-source loudspeaker (left) and a dipolar loudspeaker (right).

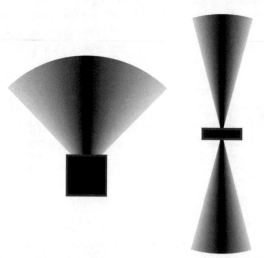

Fig.9-5 Point-source loudspeakers (left) produce sound in one direction (uni-directional radiation pattern); dipole loudspeakers (right) produce sound equally to the front and rear (dipolar radiation pattern).

Another great advantage enjoyed by ribbons is the lack of a box or cabinet. As we'll see in the section of this chapter on loudspeaker enclosures, the enclosure can greatly degrade a loudspeaker's performance. Not having to compensate for an enclosure makes it easier for a ribbon loudspeaker to achieve stunning clarity and lifelike musical timbres. Fig.9-6 is a popular full-range dipolar quasi-ribbon loudspeaker.

Fig.9-6 A full-range planar-magnetic loudspeaker. (Courtesy Magnepan)

Ribbon loudspeakers are characterized by a remarkable ability to produce extremely clean and quick transients—such as those of plucked acoustic guitar strings or percussion instruments. The sound seems to start and stop suddenly, just as one hears from live instruments. Ribbons sound vivid and immediate without being etched or excessively bright. In addition, the sound has an openness, clarity, and transparency often unmatched by dynamic drivers. Finally, the ribbon's dipolar nature produces a huge sense of space, air, and soundstage depth (provided this spatial information was captured in the recording). Some argue, however, that this sense of depth is artificially *produced* by ribbon loudspeakers, rather than being a *re*production of the actual recording.

Despite their often stunning sound quality, ribbon drivers have several disadvantages. The first is low sensitivity; it takes lots of amplifier power to drive them. Second, ribbons inherently have a very low impedance, often a fraction of an ohm. Most ribbon drivers therefore have an *impedance-matching transformer* in the crossover to present a higher impedance to the power amplifier. Design of the transformer is therefore crucial to prevent it from degrading sound quality.

Finally, some loudspeakers use a combination of dynamic and ribbon transducers to take advantage of both technologies. These so-called *hybrid* loudspeakers typically use a dynamic woofer in an enclosure to reproduce bass and a ribbon midrange/tweeter. The hybrid technique brings the advantages of ribbon drivers to a lower price level (ribbon woofers are big and expensive), and exploits the advantages of each technology while avoiding the drawbacks.

The Electrostatic Driver

Like the ribbon transducer, an electrostatic driver uses a thin membrane to make air move. But that's where the similarities end. While both dynamic and ribbon loudspeakers are electromagnetic transducers—they operate by electrically induced magnetic interaction—the electrostatic loudspeaker operates on the completely different principle of electrostatic interaction.

In the electrostatic driver, a thin moveable membrane—sometimes made of transparent Mylar—is stretched between two static elements called *stators* (Fig.9-7). The membrane is charged to a very high voltage with respect to the stators. The audio signal is applied to the stators, which create electrostatic fields around them that vary in response to the audio signal. The varying electrostatic fields generated around the stators interact with the membrane's fixed electrostatic field, pushing and pulling the membrane to produce sound. One stator pulls the membrane, the other pushes it. This illustration also shows a dynamic woofer as part of a hybrid dynamic/electrostatic system.

Electrostatic panels are of even lighter weight than planar-magnetic transducers. Unlike the ribbon driver, in which the diaphragm carries the audio signal current, the electrostatic diaphragm need not carry the audio signal. The diaphragm can therefore be very thin, often less than 0.001". Such a low mass allows the diaphragm to start and stop very quickly, improving transient response. And because the electrostatic panel is driven uniformly over its entire area, the panel is less prone to breakup. Both the electromagnetic planar loudspeaker (a ribbon) and the electrostatic planar loudspeaker enjoy the benefits of limited dispersion, which means less reflected sound arriving at the listening position. Like ribbon loudspeakers, electrostatic loudspeakers also have no enclosure to degrade the sound. Electrostatic loudspeakers also inherently have a dipolar radiation pattern. Because the diaphragm is mounted in an open panel, the electrostatic driver produces as much sound to the rear as to the front. Finally, the

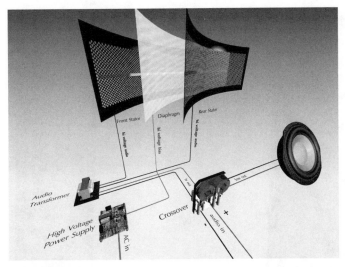

Fig.9-7 In an electrostatic loudspeaker, the moving diaphragm is a thin sheet of tightly stretched Mylar suspended between two elements called stators. This illustration shows an electrostatic panel mated to a dynamic woofer to form an electrostatic hybrid loudspeaker. (Courtesy MartinLogan)

electrostatic loudspeaker's huge surface area confers an advantage in reproducing the correct size of instrumental images.

In the debit column, electrostatic loudspeakers must be plugged into an AC outlet to generate the polarizing voltage. Because the electrostatic is naturally a dipolar radiator, room placement is more crucial to achieving good sound. The electrostatic loudspeaker needs to be placed well out into the room and away from the rear wall to realize a fully developed soundstage. Electrostatics also tend to be insensitive, requiring large power amplifiers. Nor will they play as loudly as dynamic loudspeakers; electrostatics aren't noted for their dynamic impact, power, or deep bass. Instead, electrostatics excel in transparency, delicacy, transient response, resolution of detail, stunning imaging, and overall musical coherence.

Electrostatic loudspeakers can be augmented with separate dynamic woofers or a subwoofer to extend the low-frequency response and provide some dynamic impact. Other electrostatics achieve the same result in a more convenient package: dynamic woofers in enclosures mated to the electrostatic panels. Some of these designs achieve the best qualities of both the dynamic driver and electrostatic panel. (A hybrid/dynamic electrostatic loudspeaker is illustrated in Fig.9-8.)

One great benefit of full-range ribbons and full-range electrostatics is the absence of a crossover; the diaphragm is driven by the entire audio signal. This prevents any discontinuities in the sound as different frequencies are reproduced by different drivers. In addition, removing the resistors, capacitors, and inductors found in crossovers greatly increases the full-range planar's transparency and harmonic accuracy.

Fig.9-8 An electrostatic/dynamic hybrid loudspeaker mates an electrostatic panel (top element) with a conventional dynamic woofer. (Courtesy MartinLogan)

The Dipolar Radiation Patterns of Ribbons and Electrostatics

Because planar loudspeakers (ribbons and electrostatics) are mounted in an open frame rather than an enclosed box, they radiate sound equally from the front and back. As we saw earlier, the term dipolar describes this radiation pattern. This is contrasted with point-source loudspeakers, whose drivers are mounted on the front of a box. Point-source loudspeakers are usually associated with dynamic drivers, but any type of driver in an enclosed cabinet qualifies as a point-source loudspeaker.

The dipolar radiation patterns of ribbons and electrostatics make using them very different from point-source loudspeakers. Because they launch just as much energy to the rear as they do to the front, positioning a dipole is more crucial, particularly their distance from the rear wall. Dipoles need a significant space behind them to work well. In addition, the rear wall's acoustic properties have a much greater influence over the sound. The wall behind a dipole loudspeaker should be fairly live but with some diffusion, which can be achieved with furniture or bookcases.

Horn Loudspeakers

One of the first loudspeaker technologies, the horn, has enjoyed a resurgence in popularity in the last 20 years. The horn loudspeaker employs a small dynamic driver mounted in the throat (narrow end) of a horn structure, which more efficiently couples the driver's diaphragm to the air (Fig.9-9). The horn loudspeaker operates on the same principle as a megaphone, producing increased volume. Any type of driver can be horn-loaded, but it is uncommon to find true horn-loading in woofers because of the enormous size of the horn required. The lower the frequency at which the horn is designed to operate, the larger the horn must be. Consequently, most horn-loaded loudspeakers are augmented with a conventionally loaded woofer to reproduce bass.

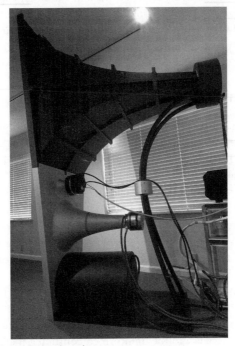

Fig.9-9 A horn-loaded system loudspeaker. (Courtesy MAGICO)

Chapter 9

Horn-loading a driver confers many performance benefits. First, horns have extraordinarily high sensitivity, and can be driven to very high volume with just a few watts of power. It's not unusual for a horn-loaded system to have a sensitivity of 100dB or more. This brings us to the second benefit of horns: they can be driven by very small power amplifiers, often those with just 10Wpc of output power. Third, the excursion (back and forth motion) of the cone or dome diaphragm can be an order of magnitude less than in a direct-radiating driver. This allows the driver to operate in the linear range of its motion at all times, greatly reducing distortion. The electrical and magnetic forces involved in moving a horn-loaded diaphragm back and forth are about one-tenth of those of a direct-radiating driver. As a result of these attributes, horn-loaded systems have state-of-the-art dynamics and lifelike recreation of music's transient signals. The sharp attack of a snare drum, for example, is reproduced with a stunning sense of realism. This quality, known as "jump factor," gives horn-loaded loudspeakers a lifelike presentation unmatched by any other loudspeaker technology.

Now the bad news. Horn-loading often introduces large tonal colorations. Try this experiment: read the previous sentence, and then read it again, this time with your hands cupped around your mouth to make a horn. For many listeners, the horn-loaded system's colorations (particularly through the midrange) are a deal-breaker, outweighing all of the horn's advantages.

There are, however, a few horn-loaded systems that do not sound colored in the slightest. These systems are very large, exotic, and extremely expensive, the result of the extraordinary manufacturing techniques required. The horn-loaded system in Fig.9-9 is an example. In this five-way system, three of the horns are machined out of solid blocks of aircraft-grade aluminum; the top horn, which handles lower-midrange frequencies, is constructed from thick aluminum reinforced with 56 hand-welded ribs.

Loudspeaker Enclosures

The enclosure in which a set of drivers is mounted has an enormous effect on the loudspeaker's reproduced sound quality. In fact, the enclosure is as important as the drive units themselves. Designers have many choices in enclosure design, all of which affect the reproduced sound.

In addition to the very different enclosure types described in this section, any speaker cabinet will vibrate slightly and change the sound. The ideal enclosure would produce no sound of its own, and not interfere with the sound produced by the drivers. Inevitably, however, some of the energy produced by the drive units puts the enclosure into motion. This enclosure vibration turns the speaker cabinet into a sound source of its own, which colors the music.

One of the factors that makes today's high-end loudspeakers so much better than mass-produced products is the extreme lengths to which high-end manufacturers go to prevent the enclosure from vibrating. Mass-market manufacturers generally skimp on the enclosure because, to the uninformed consumer, it adds very little to the product's perceived value.

A casual acquaintance once tried to impress me with how great his brother's hi-fi system was by describing how water in a glass placed on top of the loudspeaker would splash out when the system played loudly. This person held the mistaken impression that the ability to make the water fly from the glass was an indicator of the system's power and quality.

Ironically, his description told me immediately that this loudspeaker system was of poor quality. That's because low-quality loudspeaker systems have thin, vibration-prone cabinets that color the sound. The more a loudspeaker cabinet vibrates, the worse it is. Any speaker system that will splash water out of a glass *must* sound dreadful. We're about to see why.

When excited by the sound from the driver (primarily the woofer), the enclosure resonates at its natural resonant frequencies. Some of the woofer's back-and-forth motion is imparted to the cabinet. This causes the enclosure panels to move back and forth, producing sound. The enclosure thus becomes an acoustic source: we hear music not only from the drivers, but also from the enclosure.

Enclosure vibrations produce sound over a narrow band of frequencies centered on the panel's resonant frequency. The loudspeaker has greater acoustical output at that frequency. Consequently, cabinet resonances change the timbres of instruments and voices. With a recording of double-bass or piano, you can easily hear cabinet resonances as changes in timbre at certain notes.

Enclosure resonances not only color the sound spectrally (changing instrumental timbre), they smear the time relationships in music. The enclosure stores energy and releases that energy slowly over time. When the next note is sounded, the cabinet is still producing energy from the previous note. Loudspeakers with severe cabinet resonances produce smeared, blurred bass instead of a taut, quick, clean, and articulate foundation for the music. Enclosure vibration also affects the midrange by reducing clarity and smearing transient information.

The acoustic output of a vibrating surface such as a loudspeaker enclosure panel is a function of the excursion of the panel and the panel's surface area. Because

Fig.9-10 Internal cabinet bracing reduces cabinet resonances. (Courtesy Bowers & Wilkins)

loudspeaker cabinet panels are relatively large, it doesn't take much back-and-forth motion to produce sound. A large panel excited enough that you can feel it vibrating when you play music will color the sound of that loudspeaker. That's why smaller loudspeakers often sound better than similarly priced but larger loudspeakers; it's much harder (and more expensive) to keep a large cabinet from vibrating.

Cabinet vibration is why the same loudspeaker sounds different when placed on different stands, or with different coupling materials between the speaker and stand. Cones, isolation feet, and other such accessories all cause the speaker cabinet to have a different resonant signature.

A simple way of judging an enclosure's inertness is the "knuckle-rap" test. Simply knock on the enclosure and listen to the resulting sound. An enclosure relatively free of resonances will produce a dull thud; a poorly damped enclosure will generate a hollow, ringing tone.

Designers reduce enclosure resonance by constructing cabinets from thick, vibration-resistant material. Generally, the thicker the material, the better. Most loudspeakers use 3/4" Medium Density Fiberboard (MDF). MDF 1" thick is better; some manufacturers use 3/4" on the side panels and top, and 1" or 2" MDF for the front baffle, which is more prone to vibration. Exotic materials and construction techniques are also used to combat cabinet resonances. Some manufacturers go to extreme lengths to keep their cabinets inert, including advanced materials research and extraordinary construction techniques.

Cross-braces inside the enclosure reduce the area of unsupported panels and make the cabinet more rigid. Some braces are called "figure-8" braces because their four large holes form a figure-8 pattern. These braces are strategically placed for maximum effectiveness. (Heavy-duty internal cabinet bracing is shown in Fig.9-10.)

Most quality loudspeakers are equipped with threaded inserts on the bottom panel to accept spikes. These spikes better couple the speaker to the floor and reduce enclosure vibration.

Similarly, small loudspeakers should be mounted on solidly made stands for best performance. In fact, the stand's quality can greatly affect the reproduced sound. Flimsy, lightweight stands should be avoided in favor of rigid models. The stand should include spikes on its bottom plate to better couple the stand and loudspeaker to the floor. Some loudspeaker stands can be filled with sand or lead shot for *mass loading*, making them more inert and less prone to vibration. A great loudspeaker on a poor stand will suffer significantly degraded performance. Plan to spend several hundred dollars on stands. When comparing a floorstanding loudspeaker to one requiring stand-mounting, include the cost of the stands in your budgeting.

The interface between loudspeaker and stand also deserves attention. Spikes, cones, and other isolation devices (see Chapter 11) can allow the loudspeaker to perform at its best. An effective yet inexpensive interface is a sticky, gum-like material called Bostik Blue-Tak, available at hardware stores. Simply place a small ball of Blue-Tak at each corner of the speaker stand.

Enclosure Shapes

The enclosure can also degrade a loudspeaker's performance by creating diffraction from the cabinet edges, grille frame, and even the drivers' mounting bolts. Diffraction is a re-radiation of energy when the sound encounters a discontinuity in the cabinet, such as at the enclosure edge. Diffracted energy combines with the direct sound to produce ripples (i.e., colorations) in the frequency response. Rounded baffles, recessed drivers and mounting bolts, and low-profile grille frames all help to reduce diffraction.

Some loudspeaker enclosures are tilted back to align the drivers in time. This tilt aligns the acoustic centers of all the drivers so that their outputs arrive at the listener at the same time (Fig.9-11).

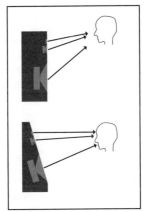

Fig.9-11 An angled baffle helps to time-align the loudspeaker's individual drive units at the listening position.

Air-Suspension and Bass-Reflex Enclosures

The *air suspension* enclosure (technically called an *infinite baffle*) simply seals the enclosure around the driver rear to prevent the two waves from meeting. The air inside the sealed enclosure acts as a spring, compressing when the woofer moves in and creating some resistance to woofer motion.

I mentioned earlier that a woofer in a box produces just as much sound inside the box as outside. The *bass-reflex* enclosure exploits this fact, channeling some of this bass energy out of the cabinet and into the listening room. You can instantly spot a reflex-loaded loudspeaker by the hole, or vent, on the front or back of the enclosure. Reflex-loaded loudspeakers are also called *ported* designs.

Reflex loading extends the low-frequency cutoff point by venting some of the energy inside the enclosure to the outside. That is, a reflex-loaded woofer will go lower in the bass than one in a sealed enclosure, other factors being equal.

Reflex loading has three main advantages. First, it increases a loudspeaker's maximum acoustic output level—it will play louder. Second, it can make a loudspeaker more sensitive—it needs less amplifier power to achieve the same volume. Third, it can lower a loudspeaker's cutoff frequency—the bass goes deeper. Note that these benefits are not available simultaneously; the acoustic gain provided by reflex loading can be used either to increase a loudspeaker's sensitivity or to extend its cutoff frequency, but not both.

On the down side, a woofer in a vented enclosure will tend to keep moving after the drive signal has stopped. This difference in transient response can be manifested as sluggish sound in, for example, a kickdrum. A sealed enclosure has better transient response and better bass definition, but at the cost of lower sensitivity and less deep bass extension. Nonetheless, many very-high-quality loudspeakers have been based on ported designs.

Passive Radiators

A variation on the reflex system is the passive radiator, also called an auxiliary bass radiator, or ABR. This is usually a flat diaphragm with no voice coil or magnet structure that cannot produce sound on its own. Instead, the diaphragm covers what would have been the port in the reflex system, and moves in response to varying air pressure inside the cabinet caused by the woofer's motion.

Powered and Servo-Driven Woofers

Loudspeaker designers are increasingly choosing to include a power amplifier within the loudspeaker to drive the woofer cone. In such designs, the external power amplifier is relieved of the burden of driving the woofer, and the loudspeaker designer has more control over how the woofer behaves. For example, equalization can be applied to extend the bass response.

Some designers don't try to juggle these laws of physics to produce the most musically satisfying compromise in bass performance; instead, they take brute-force control of woofer movement with the servo-driven woofer. A servo-woofer system consists of a woofer with an accelerometer attached to the voice coil, and a dedicated woofer power amplifier. An accelerometer is a device that converts motion into an electrical signal. The accelerometer sends a signal back to the woofer amplifier, telling the woofer amplifier how the woofer cone is moving. The woofer amplifier compares the drive signal to the cone's motion; any difference is a form of distortion. The woofer amplifier can then change the signal driving the woofer so that the woofer cone behaves optimally.

Crossovers

A loudspeaker crossover is an electronic circuit inside the loudspeaker that separates the frequency spectrum into different ranges and sends each frequency range to the appropriate drive unit: bass to the woofer, midband frequencies to the midrange, and treble to the tweeter (in a three-way loudspeaker). Fig.9-12 illustrates this process.

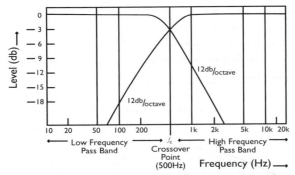

Fig.9-12 A loudspeaker's crossover splits up the frequency spectrum, sending bass to the woofer, midrange frequencies to the midrange driver and treble to the tweeter.

A crossover is made up of capacitors, inductors, and resistors. These elements selectively filter the full-bandwidth signal driving the loudspeaker, creating the appropriate filter characteristics for the particular drivers used in the loudspeaker. A crossover is usually mounted on the loudspeaker's inside rear panel.

A crossover is described by its *cutoff frequency* and *slope*. The cutoff frequency is the frequency at which the transition from one drive unit to the next occurs—between the woofer and midrange, for example. The crossover's slope refers to the rolloff's steepness. A slope's steepness describes how rapidly the response is attenuated above or below the cutoff frequency. For example, a *first-order crossover* has a slope of 6dB/octave, meaning that the signal to the drive unit is halved (a reduction of 6dB) one octave above the cutoff frequency. If the woofer crossover circuit produces a cutoff frequency of 1kHz, the signal will be rolled off (attenuated, or reduced in level) by 6dB one octave higher, at 2kHz. In other words, the woofer will receive energy at 2kHz, but that energy will be reduced in level by 6dB. A first-order filter producing a 6dB/octave slope is the most gentle rolloff used.

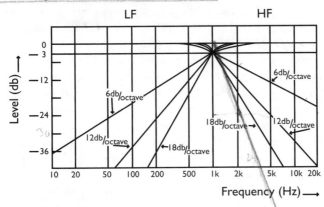

Fig.9-13 Crossovers are specified by their slope, or how steeply they attenuate frequencies above or below their cutoff frequency.

The next-steeper filter is the *second-order crossover,* which produces a rolloff of 12dB/octave. Using the preceding example, a woofer crossed over at 1kHz would still receive energy at 2kHz, but that energy would be reduced by 12dB (one quarter the amplitude) at 2kHz. A *third-order crossover* has a slope of 18dB/octave, and a *fourth-order crossover* produces the very steep slope of 24dB/octave. Using the previous example, the fourth-order filter would still pass 2kHz to the woofer, but the amplitude would be down by 24dB (1/16th the amplitude). Fig.9-13 compares crossover slopes.

Typical crossover points for a two-way loudspeaker are between 1kHz and 2.5kHz. A three-way system may have crossover frequencies of 800Hz and 3kHz. The woofer reproduces frequencies up to 800Hz, the midrange driver handles the band between 800Hz and 3kHz, and the tweeter reproduces frequencies above 3kHz.

Digital Loudspeakers

The capacitors, inductors, and resistors that make up a loudspeaker crossover are far from perfect. Not only do they exhibit variations in value that affect performance, these components split up the frequency spectrum in a relatively crude way. Moreover, a significant amount of amplifier power is wasted in the crossover.

The widespread availability of digital-audio sources provides an opportunity to remove traditional crossovers from the signal path. If we can divide the frequency spectrum digitally—that is, by performing mathematical computations on the digital representation of the audio signal—we can create just about any crossover characteristics we want, with none of the problems of capacitors, inductors, and resistors.

These "digital loudspeakers" accept a digital input signal and implement the crossover in the digital domain with *digital signal-processing (DSP)* chips. Instead of subjecting the high-powered audio signal to resistors, capacitors, and inductors, DSP crossovers separate the frequency spectrum by performing mathematical processing on the digital audio data. DSP crossovers can be programmed to have perfect time behavior, as well as use any slope and frequency the designer wants without regard for component limitations or tolerances, and employ equalization to the individual drive units.

Fig.9-14 is a side view of a combination pictorial/block diagram of a digital loudspeaker. The speaker accepts a digital input signal from a CD transport, DVD-Audio player, or other digital source. DSP chips inside the speaker split the frequency spectrum into bass, midrange, and treble. These chips can also equalize and delay signals to produce nearly perfect acoustic behavior at the drive units' output. Each of the three digital signals (bass, midrange, treble) is then converted into an analog signal with its own digital-to-analog converter (DAC). Each DAC's output feeds a power amplifier specially designed to power the particular drive unit used in the digital loudspeaker. This aspect of a digital loudspeaker confers a large advantage: The power amplifiers are designed to drive a known load. The power amplifiers amplify an analog signal, just like a conventional power amp, and then drive conventional cone loudspeakers.

In addition to employing virtually perfect crossovers, digital loudspeakers also simplify your system. You need only a digital source and a pair of loudspeakers for a

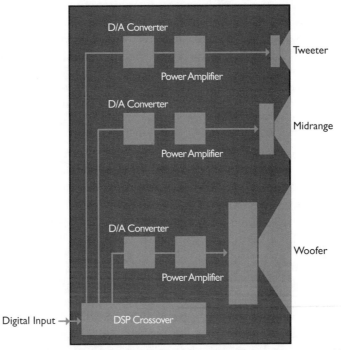

Fig.9-14 Digital loudspeakers employ a digital-to-analog converter and power amplifier for each driver.

complete music system. This eliminates large power amplifiers and garden-hose speaker cable on your floor. Moreover, the digital nature of the system provides for remote control of the loudspeaker. You can adjust the tonal balance, correct for different listening heights, and even delay one loudspeaker if your room doesn't permit symmetrical speaker placement. The digital speaker will also have a volume readout, which makes setting the correct playback level for different music more convenient. Another advantage of a loudspeaker under DSP control is the ability to protect the drivers from damage; the DSP knows the driver's limitations and can impose a maximum excursion limit. When considering the price of digital loudspeakers, keep in mind that you don't need to buy an outboard digital-to-analog converter or power amplifiers.

Digital loudspeakers were pioneered by Meridian Audio, which has been producing them since 1993. Meridian's first models accepted the 44.1kHz, 16-bit datastream from a CD transport, but newer models have been upgraded to accept high-resolution digital audio with sampling rates up to 192kHz and word lengths up to 24 bits.

Subwoofers

A *subwoofer* is a loudspeaker that produces low frequencies that augment and extend the bass output of a full-range loudspeaker system. The term "subwoofer" is grossly misused to describe any low-frequency driver system enclosed in a separate cabinet. But "subwoofer" actually means "below the woofer," and should be reserved for those products that extend bass response to below 20Hz. A low-frequency driver in an enclosure extending to 40Hz and used with small satellite speakers is more properly called a woofer.

You'll also see full-range speakers with a built-in "subwoofer" powered by its own amplifier. Most of these products actually employ woofers that are simply driven by an integral power amplifier. Such a design relieves your main amplifier of the burden of driving the woofer, but requires that the loudspeakers be plugged into an AC outlet.

Subwoofers come in two varieties: *passive* and *active*. A passive subwoofer is just a woofer or woofers in an enclosure that must be driven by an external amplifier. An active subwoofer combines a subwoofer with a line-level crossover and power amp in one cabinet. Such a subwoofer has line-level inputs (which are fed from the preamplifier), line-level outputs (which drive the power amp), and a volume control for the subwoofer level. In some active subwoofers, an integral crossover separates the bass from the signal driving the main loudspeakers at line-level, which is much less harmful to the signal than speaker-level filtering. The line-level output is filtered to roll off low-frequency energy to the main loudspeakers. This crossover frequency is often adjustable on subwoofers to allow you to select the frequency that provides the best integration with the main loudspeakers (more on this later).

A subwoofer used as part of a home-theater loudspeaker system is often connected in a completely different manner. When used for home-theater, the subwoofer is driven by the SUBWOOFER OUTPUT jack of your A/V receiver or A/V controller. In this case, the receiver or controller performs the crossover function, sending only bass to the subwoofer. Some subwoofers have separate inputs for home-theater connection (the input is often labeled "Processor") and conventional stereo reproduction (the input is often labeled "Line"). (Chapter 10 includes a more complete description of these connection choices.)

Adding an actively powered subwoofer to your system can greatly increase its dynamic range, bass extension, midrange clarity, and ability to play louder without strain. The additional amplifier power and low-frequency driver allow the system to reproduce musical peaks at higher levels. Moreover, removing low frequencies from the signal driving the main loudspeakers lets the main loudspeakers play louder because they don't have to reproduce low frequencies. The midrange often becomes clearer because the woofer cone isn't furiously moving back and forth trying to reproduce low bass.

A surprising additional benefit of adding a subwoofer is an increased sense of space and soundstage size, even when playing music with little low-frequency content. I've heard a demonstration of an unaccompanied singer in a large hall with a subwoofer turned on and turned off. One would expect that a solo voice would not be affected by the presence of a subwoofer, but with the sub, the hall size appeared to increase. The reason is that low-frequencies contain subtle spatial cues (in recordings made in large rooms) about the size of the acoustic space.

Unfortunately, many subwoofers degrade a playback system's musical performance. Either the subwoofer is poorly engineered (many are), set up incorrectly, or, as is increasingly common, designed for reproducing explosions in a home-theater system, not resolving musical subtleties. Another problem is getting the subwoofer to *integrate* with the main speakers so that you hear a seamless and coherent musical whole. As instruments traverse the crossover region, you should hear no discontinuity in the sound. Ascending and descending bass lines, for example, should flow past the crossover point with no perceptible change in timbre or dynamics.

All of these problems are exacerbated by most people's tendency to set subwoofer levels way too high. The reasoning is that if you pay good money for something, you want to hear what it does. But if you're *aware* of the subwoofer's presence, either its level is set too high, it isn't positioned correctly, or the subwoofer has been poorly designed. The highest compliment one can pay a subwoofer is that its contribution can't be heard directly. It should blend seamlessly into the musical fabric, not call attention to itself.

For home theater, a subwoofer is essential to the movie-watching experience. A good subwoofer adds an excitement and visceral thrill to some movies. If you have full-range left and right speakers that have adequate bass response for music, drive the subwoofer with the receiver or controller's subwoofer output signal. In the receiver or controller's set-up menu, select LARGE for the left and right speakers. With this arrangement, the subwoofer won't be engaged when playing stereo music through the left and right loudspeakers.

Chapter 10 includes some special set-up techniques for connecting a subwoofer and getting it to sound good in your system.

10

Audio for Home Theater and Multichannel Music

The explosive growth of home theater and high-definition television has expanded the possibilities for a home-audio system. Today's technology allows us to enjoy 2-channel music, multichannel music, and film soundtracks in surround sound in our homes—all through the same audio system.

In this chapter we'll look at the components needed for multichannel-music and home-theater reproduction, as well as ways of enjoying home theater without compromising the system's musical performance. (For a more in-depth treatment of home theater, including video components, see *Home Theater for Everyone*, published by Acapella Publishing.)

Overview—Surround Sound

Home theater is made possible by multichannel surround sound. Instead of two speakers in front of you, as in a stereo system, home theater surrounds you with five, six, or even seven. Three speakers are placed in the front of the room, with two, three, or four surround speakers located to the rear or sides of the listening/viewing position. Surround sound is what provides the sense of envelopment, of being in the action, that makes the home-theater experience so engrossing.

Multichannel surround sound is encoded on DVDs, some satellite and cable TV transmissions, and over-the-air high-definition television (HDTV) broadcasts in the Dolby Digital format. This format, introduced in 1997 to replace Dolby Surround (and its decoding variant, Dolby Pro Logic), encodes six discrete audio channels in a single digital bitstream. The six channels are called *left, center, right, left surround, right surround*, and *low-frequency effects* (LFE), and correspond to loudspeakers arrayed in the pattern shown in Fig.10-1. The LFE channel carries only high-impact, low-bass information (below 100Hz), and can be reproduced by any of the loudspeakers, or by a separate subwoofer. Typically, the LFE channel drives a subwoofer. With five full-bandwidth channels plus the 100Hz LFE channel, we call Dolby Digital a *5.1-channel* format. The ".1" in 5.1-channel sound is the LFE channel. The competing DTS surround-sound format, found on some DVDs as an option, also delivers 5.1 discrete audio channels to your home-theater system.

Home-theater audio systems are increasingly moving from 5.1 channels to 7.1 channels. Many audio/video receivers and controllers can accommodate 7.1 channels. The two additional speakers in a 7.1-channel system, called *surround-back* speakers, are mounted on the rear wall directly behind the listener. This loudspeaker array takes advantage of the newer surround-sound formats that deliver an additional signal to drive the surround-back speakers. In addition, A/V receivers and controllers with 7.1-channel capability can synthesize signals to drive the surround-back speakers from 5.1-channel sources such as Dolby Digital and DTS.

(Note that some receivers have six amplification channels; in that setup, only one surround-back speaker is employed directly behind the listener. This configuration is, however, uncommon; if you're going to add surround-back speakers, it's not much additional effort or cost to add two of them.)

Fig.10-1 A home-theater audio system typically employs three speakers across the front of the room, two surround speakers to the room sides or rear, and a subwoofer. (Courtesy Bowers and Wilkins)

Should You Choose a 5.1-Channel or a 7.1-Channel System?

A fundamental question facing the home-theater shopper is whether to buy a 5.1- or 7.1-channel audio system. The vast majority of today's film soundtracks are encoded in 5.1 channels, but newer films may be mixed in 7.1 channels for theatrical exhibition as well as for home theater. The two new high-definition disc formats vying to replace DVD (Blu-ray Disc and HD DVD) can carry 7.1-channel soundtracks. The future might be 7.1, but the present is rooted in 5.1 channels.

There are several arguments for choosing a 5.1-channel system. First, most films are mixed in 5.1 channels. Second, mounting and running cables to the additional surround-back speakers isn't worth the small improvement in performance (except in very large rooms). Third, many extremely satisfying home-theater systems are based on 5.1-channel playback. Fourth, a 7.1-channel loudspeaker package distributes your speaker expenditure over seven speakers plus a subwoofer rather than over five speakers and the sub, which could compromise the quality—two good surround speakers are better than four mediocre ones. Finally, if your living room dictates that your couch be positioned against the wall opposite the video display, there's little point in adding surround-back speakers on that wall; it's not likely that they will perform as intended.

Audio for Home Theater and Multichannel Music

The argument for purchasing a 7.1-channel system goes like this: Even though most films are mixed in 5.1 channels, today's A/V receivers and controllers have processing circuits to create the surround-back channels from 5.1-channel sources. Although this trick doesn't provide true 7.1-channel playback (the surround-back signals are simply derived from the left- and right-surround channels), it nonetheless increases the feeling of immersion in the soundfield. When playing true 7.1-channel sources, the film-sound mixers are able to position sounds directly behind you, an impossible feat with just 5.1 channels. Moreover, we are moving into an increasingly 7.1-channel world with the new audio formats on HD DVD and Blu-ray Disc; it's better to get a 7.1-channel system now rather than have to upgrade later. (See the Addendum at the end of this chapter for more detail on surround-sound formats.)

Here's my view. If you have a moderately sized room and you're adding surround speakers for the first time, go with a 5.1-channel system. If your room is large, or if you're building a theater room as part of new construction, select a 7.1-channel package. Larger rooms benefit from the surround-back speakers because they help to fill the "hole" between widely spaced left- and right-surround speakers. If you're building a home with surround-speaker wiring in the walls, there's very little cost and effort required to run wires for surround-back speakers.

Overview—Audio/Video Receivers, Controllers, and Amplifiers

The job of decoding the Dolby Digital or DTS bitstream into separate audio channels, and then converting those digital data to analog signals, falls to the audio/video receiver (AVR) or A/V controller (the latter is also known as a *surround-sound processor*). The difference between them is that the AVR contains five, six, or seven amplifier channels; the controller has no internal amplification and must be used with a separate multichannel amplifier. The AVR also includes an AM/FM tuner and possibly XM or Sirius satellite radio. Other than those differences, the two components are identical.

Let's first look at the A/V controller. In addition to performing surround-sound decoding (Dolby Digital, DTS, Pro Logic, etc.), the controller receives audio and video signals from the source components (satellite, DVD, tuner, CD, SACD, DVD-A, digital-video recorder) and selects which one is decoded and amplified by your home-theater audio system and sent to your video monitor for display. The controller also performs digital signal processing and bass management, adjusts the overall volume, and fine-tunes the levels for individual channels (these processes are explained later in this chapter). The controller's output is six line-level signals: the left, center, right, left-surround, right-surround, and subwoofer channels. Controllers that support the newer 7.1-channel surround-sound formats (Dolby Digital EX, DTS-ES, Dolby Digital Plus, DTS-HD) have eight line-level outputs rather than six. The eight outputs correspond to seven audio channels plus a subwoofer.

These six or eight separate outputs are fed to a 5- or 7-channel power amplifier (plus an optional subwoofer), where they are amplified to a level sufficient to drive the home theater's loudspeaker system. A home-theater power amplifier can have five, six, or seven amplifier channels in a single chassis. Alternately, some power amplifiers have three channels, others have two, and some are available as monoblocks: one amplifier channel per chassis. A 3-channel power amplifier allows you to use your existing 2-channel stereo power amplifier on the surround channels and the 3-channel amplifier on the front three loudspeakers, for example. Most subwoofers have built-in power amplifiers to drive their woofer cones.

An audio/video receiver (AVR) incorporates, in one chassis, an A/V controller and multichannel power amplifier. AVRs are less expensive than separate controllers and power amplifiers. Separate controllers and power amplifiers usually deliver better performance than an AVR, although today's upper-end AVRs are quite sophisticated, and many offer superb sound quality.

Overview—Home-Theater Loudspeakers

A conventional stereo system provides two audio channels, left and right, which are reproduced by the left and right loudspeakers. When correctly set up, two channels of information driving two loudspeakers produce a soundfield in front of the listener that seems to exist between and around the two loudspeakers.

A home-theater system provides multiple channels, each channel feeding a loudspeaker located in front of, alongside, and/or behind the listening/viewing position. Specifically, a home-theater system has three *front* loudspeakers located across the front of the room, and two, three, or four surround loudspeakers behind or to the side of the listening position. Two of the front three loudspeakers flank the video monitor; the third is positioned above or below the video monitor. The front loudspeakers are called *left*, *center*, and *right* (Fig.10-1 earlier).

The left and right loudspeakers reproduce mostly music and sound effects. The center loudspeaker's main job is reproducing dialogue, and anchoring onscreen sound effects on the television screen. Having three loudspeakers across the front of the room helps the sound's location more closely match the location of the sound source in the picture. For example, in a properly set-up home-theater system, if a car crosses from the left side of the picture to the right, you hear the sound of the car move from the left loudspeaker, through the center loudspeaker, and then to the right loudspeaker. The sound source appears to follow the image on the screen.

The surround loudspeakers have a different job. They're generally smaller than the front loudspeakers, and handle much less energy. Consequently, they can be mounted unobtrusively on or inside a wall. Surround speakers mostly reproduce "atmospheric" or ambient sounds, creating a diffuse aural atmosphere around the listener. In a jungle scene, for example, the surround loudspeakers would re-create sounds such as chirping birds, falling raindrops, and blowing wind; in a city scene, the viewer would be surrounded by traffic sounds. The surround loudspeakers' contribution is subtle, but vitally important to the overall experience. Correctly set-up surround loudspeakers should not be able to be heard directly, but should instead envelop the viewer/listener in a diffuse soundfield.

Let's look at each of these components of a home-theater system in more detail.

A/V Receivers

The audio/video receiver is a marvel of technological sophistication, packing advanced features and technologies into a chassis about the size of yesterday's stereo receiver. The following discussion of AVRs applies equally well to A/V controllers, with the exception that controllers have no built-in amplification or radio reception.

A modern A/V receiver, shown in Fig.10-2, performs five vital functions in a home-theater system: the source-switching functions of a preamplifier (selecting which source to listen to or watch), surround-sound decoding, six (or eight) channels of digital-to-analog conversion, an electronic crossover to split up the frequency spectrum, and five, six, or seven channels of amplification.

Fig.10-2 A modern AVR is packed with sophisticated digital signal processing. (Courtesy Rotel)

Specifically, an AVR performs these functions:

1) receives video and audio signals from various source components (digital-video recorder, satellite, cable, DVD, HD DVD, Blu-ray Disc, etc.) and selects which are sent to the video monitor and home-theater audio system (this function is called "source switching");

2) performs surround decoding, whether Dolby Pro Logic, Pro Logic II, Pro Logic IIx, Dolby Digital, DTS, Dolby Digital EX, DTS-ES, or one of the newer high-resolution surround-sound formats (the Addendum to this chapter explains the many surround-sound formats and their variations);

3) controls playback volume;

4) makes adjustments in system setup, such as the individual channel levels;

5) directs bass to the appropriate loudspeakers or subwoofers (a function called *bass management*);

6) performs THX processing (if so equipped);

7) amplifies the five, six, or seven channels to a level sufficient to drive a loudspeaker array.

The AVR is the master control center of your home-theater system. When you first set up an AVR, its onscreen display will guide you through selecting such things as whether or not you're using a subwoofer, if your surround loudspeakers are large or small, and in calibrating individual channel levels and delay times (Fig.10-3). You will also control the system on a day-to-day basis through the AVR's front panel and remote control. This includes source selection, volume adjustment, and the setting of individual channel levels (described later in this chapter).

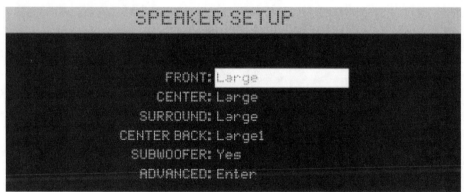

Fig.10-3 Many of an AVR's set-up parameters are accessed via on on-screen display. (Courtesy Rotel)

How to Choose a Receiver

Receivers range from entry-level units costing about $200 to state-of-the-art models that will set you back as much as $8000. This wide price range reflects the significant differences between receivers in their features, connectivity, power outputs, and sonic performances.

When choosing a receiver, first make sure the receiver has enough inputs for your system, provides your preferred surround-format decoding (Dolby Digital EX and DTS-ES, for example), offers a friendly user interface, has THX processing and certification (if desired), is well-built, and has good sound quality.

Inputs, Outputs, and Source Switching

Let's start with the receiver's most basic function: selecting the source you listen to or watch. The receiver accepts audio or A/V (audio and video) signals from all your source components and lets you select which source signal is sent to the power amplifiers and video monitor. A basic receiver will offer two analog-audio inputs and perhaps four audio/video (A/V) inputs. Most receivers today offer one USB input for connecting your computer to play music files from your hard disk. In addition to the main outputs that drive your TV and power amplifiers, two *record outputs* are often provided, to drive two video recorders.

When choosing a receiver, make sure its array of inputs matches or exceeds the number of source components in your system. Your system is likely to expand in the future, so look for a receiver with at least one more input than you need right now.

All receivers have inputs for digital audio signals as well as for analog ones. These inputs receive the digital-audio output of a DVD player, satellite receiver, digital-video recorder, or other digital source. This digital connection can transmit a variety of digital signals, including Dolby Digital, DTS, Dolby Surround, and 2-channel PCM (Pulse Code Modulation) signals. The latter signal appears on a CD player's digital-output jack, or a DVD player's digital-output jack when playing CDs. As just mentioned, a USB input is handy for playing files from your computer through your main audio system. Some AVRs also have a port for connecting an iPod. A few offer a signal-processing mode that reportedly improves the sound of compressed MP3 files to "CD quality" (a doubtful claim). Also consider the type of video jacks on the A/V inputs. Most receivers offer S-video jacks on all inputs, with perhaps two inputs with component-video connections. Component video, carried on three RCA jacks, delivers better performance than S-video. If you plan to connect multiple sources via component video, be sure the receiver can accommodate all your component-video sources.

As more source components and video monitors incorporate High-Definition Multimedia Interface (HDMI) inputs, receivers are now offering HDMI switching. That is, the receiver may have several HDMI inputs and one HDMI output, allowing you to select which HDMI-equipped source is displayed on your video monitor. If your receiver lacks this feature, you'll need to run the HDMI output from your high-definition source components directly to the HDMI inputs of your video monitor. Some video displays have only one or two HDMI inputs; if you have more HDMI-equipped source components than HDMI inputs on your video display, you'll need HDMI source switching. Fig.10-4 shows the rear panel of a controller equipped with HDMI switching, and an HDMI connector is shown on page 54.

Fig.10-4 Today's AVRs and controllers provide HDMI switching, seen here as the top right row of jacks. (Courtesy Anthem)

All AVRs have a video output that connects to your video display, with the receiver selecting which video input feeds the display. Some AVRs have an extremely useful feature called *upconversion* or *transcoding* (the terms describe the same feature) that converts all video sources (composite video, S-video, component video, or HDMI) to an HDMI output. This allows you to run just a single HDMI cable between your receiver and video display. Without this feature, you would need to connect as many types of video cables as you had different outputs on your video sources. For example, if you had a DVD player with component-video output and a satellite receiver with S-video, you would need to run both component-video and S-video cables between the receiver and television. HDMI upconversion solves this problem.

Digital Signal Processing (DSP) and Surround-Decoding Formats

The availability of powerful Digital Signal Processing (DSP) chips has revolutionized receivers in the past few years. A DSP chip is a number cruncher that operates on specific instructions (the software) controlling it. DSP chips are the heart and brain of the receiver, performing surround-sound decoding, signal processing (equalization, crossovers), and THX post-processing (if the receiver is THX-certified). Today's advanced receivers boast the computing power of a late-1980s mainframe computer.

The first job of the DSP chip is decoding, that is, converting a stream of digital data into separate digital signals that can be converted to analog audio. All receivers today decode the three major surround-sound formats: Dolby Digital, DTS, and Dolby Surround. Dolby Digital is the mandatory surround-sound format for DVD (DTS is optional), and has been chosen as the surround-sound format for high-definition television (HDTV). Dolby Digital Plus and DTS-HD decoding are required in in HD DVD and Blu-ray disc players, with studios free to choose either format. Dolby TrueHD and DTS-HD Master Audio are optional on HD DVD and Blue-ray discs.

6-Channel Analog Input

To play multichannel music, the receiver must have a six-channel (or eight-channel) analog input. This input is fed from the six analog outputs from a multichannel SACD or DVD-Audio player. (See page 53 for a complete explanation.)

Analog-Bypass Mode

For the audiophile shopping for a receiver that will also serve as a 2-channel preamplifier for his system, one of the most significant considerations is its performance with 2-channel analog sources and signals from an SACD or DVD-Audio player. If uncompromised music performance is important to you, you'll want a receiver that has an *analog-bypass mode*. Without an analog-bypass mode, all analog signals will be converted to digital and back to analog as they pass through the receiver. Digital conversion is far from sonically transparent, so the sound will suffer.

Audio for Home Theater and Multichannel Music

Automatic Calibration

Some receivers offer automatic calibration of individual channel levels and delay times. This calibration is covered in detail later in this chapter. Briefly, home-theater systems require that you adjust the level of each channel so that the listener hears the same volume from each speaker. Similarly, you must tell the AVR the distance between each loudspeaker and the listening positions; the receiver delays the signal from some channels so that the sound from each speaker arrives at the listening position at the same time. Automatic calibration in AVRs makes this process simple and foolproof; you position a measurement microphone at your listening seat, press a few buttons, and the system calibrates itself. This feature, once reserved for very expensive AVRs and controllers, is now found even in budget models.

7.1-Channel Playback from 5.1-Channel Sources

Some home-theater controllers and receivers provide 7.1-channel playback from 5.1-channel sources such as Dolby Digital and DTS. These products have eight channels (seven channels plus a subwoofer) rather than the six channels of most Dolby Digital- and DTS-equipped products. The two additional channels drive two extra speakers placed behind the listener, augmenting the two surround speakers at the sides of the listening location.

Making effective use of the additional surround speakers requires sophisticated signal processing in the receiver. Receivers with THX Ultra2 processing incorporate a circuit that creates a 7.1-channel signal from any source, allowing you to use your 7.1-channel loudspeaker system on all source material, not just on those movies that have been encoded with 7.1 channels of information in the Dolby Digital Plus, Dolby TrueHD, DTS-HD, or DTS-HD Master Audio formats (see the end of this chapter). Dolby's Pro Logic IIx will also generate from 5.1-channel sources a 7.1-channel signal that can drive the eight loudspeakers in some systems.

Multi-Room and Multi-Zone

Some upper-end AVRs offer a feature called *multi-room* or *multi-zone*. Both terms describe the receiver's ability to send a line-level signal to an amplifier and speakers in another room. This allows one person in the family to watch a movie in the living room while another listens to XM or Sirius satellite radio in the bedroom, for example. If you see a button on the front panel marked ZONE 2, the receiver has multi-room capability.

Power Output in AVRs

AVRs vary in their power outputs from a low of about 60Wpc to a whopping 200Wpc. When choosing an AVR, match the AVR's output power to the sensitivity of the loudspeakers it will be asked to drive (or vice versa). The criteria used in matching an amplifier's output power to a loudspeaker's sensitivity, described in detail in Chapters 8 and 9, apply equally to an AVR and a multichannel speaker system.

Note, however, that AVRs generally don't perform as well as separate amplifiers in their ability to drive loudspeakers. The designers must squeeze into a chassis of manageable size the power supplies, output transistors, and heat sinks—all large and heavy items—for five, six, or seven channels. The power supply is what delivers electrical current to the receiver's output transistors—current that ultimately pushes and pulls the cones in your loudspeakers back and forth. Consequently, the first performance attribute sacrificed in inexpensive products is the ability to deliver current when the loudspeaker demands it. In fact, some receivers will overheat and shut down if asked to drive loudspeakers with an impedance of less than 6 ohms. The receiver's output transistors simply aren't up to the job. (A more thorough explanation of this phenomenon is included in Chapter 8.) This rarely occurs in separate power amplifiers because of their greater capacity to deliver current to the loudspeakers.

A great solution is to match the AVR with high-sensitivity speakers rated at 8 ohms. Remember that choosing a speaker system of just 3dB more sensitivity than another is equivalent to doubling the amplifier power. Put another way, a speaker rated at 90dB sensitivity puts *half* as much demand on the receiver as a speaker rated at 87dB sensitivity.

Satellite Radio: XM and Sirius

Many of today's AVRs are equipped with XM Satellite Radio or Sirius Satellite Radio capability (XM is more prevalent in AVRs). Both services offer more than 100 commercial-free channels of music, news, sports, talk, and other programming. (XM offers 70 music channels, 30 non-music channels; Sirius delivers 125 channels some of which are non-music channels such as sports, weather, and talk radio.) Unlike over-the-air radio which is free to everyone, you must purchase a subscription to XM or Sirius. XM currently charges $9.95 per month, Sirius $12.95. You must also purchase an XM or Sirius antenna to connect to the AVR.

The big attraction of XM and Sirius, aside from a wide selection of commercial-free music, is the ability to receive this programming anywhere in the country with no static, fading, or the other problems associated with analog broadcasting. If you've become hooked on XM or Sirius in your car, look for an AVR with XM or Sirius capability.

FM Tuner Performance

If you plan to use your AVR to receive FM broadcasts, consider the receiver's tuner specifications before purchasing. Although we can't tell much about how an AVR will sound by looking at the specification sheet, FM tuner performance can be characterized by a few key specs.

Good tuners are characterized by their *sensitivity*, or ability to pull in weak stations. The greater a tuner's sensitivity, the better it can pick up weak or distant stations. This aspect of a tuner's performance is more important in suburban or rural areas that are far from radio transmitters.

A tuner characteristic of greater importance to the city dweller is *adjacent-channel selectivity*—the ability to pick up one station without interference from the station next to it on the dial. The *alternate-channel selectivity* specification defines a tuner's ability to reject a strong station two channels away from the desired channel. When stations are packed closely together, as they are in cities, adjacent-channel and alternate-channel selectivity are more important than sensitivity.

Equally important to all listeners is the tuner's *signal-to-noise ratio*, a measure of the difference in dB between background noise and the maximum signal strength. A tuner with a poor signal-to-noise ratio will overlay the music with an annoying background hiss.

In short, a poor tuner will have trouble receiving weak stations, may lack the ability to select one station when that station is adjacent to another station, have high background noise, and be overloaded by nearby FM transmitters or other radio signal sources (taxi dispatchers, for example).

THX Certification of AVRs and Controllers

Some AVRs are THX-certified, meaning that they incorporate certain types of signal processing to better translate film sound into a home environment as well as meet a set of specific technical criteria established by Lucasfilm. THX processing in an AVR includes *re-equalization, surround decorrelation*, and *timbre matching*. Re-equalization removes excessive brightness from film soundtracks for home-theater playback. Surround decorrelation improves the sense of spaciousness from the surround channels by introducing slight differences between the two surround signals. Timbre matching compensates for the fact that the ear doesn't hear the same tonal balance of sounds from all directions. The THX timbre-matching circuit ensures that sounds from the front and rear have the same timbre (tonal balance). THX-certified AVRs must also meet certain power-output requirements. All THX-certified AVRs also have a standardized subwoofer crossover circuit.

There are various levels of THX certification. AVRs with THX-Select certification contain all the signal processing described above, but their power-output requirements are relaxed so that THX processing can be made available in less expensive AVRs that are suitable for smaller rooms. What was the original THX specification is now called THX Ultra. A third THX certification, called THX Ultra2, applies some new criteria to the certification requirements. First, the receiver (or controller) must have seven channels (rather than five or six); signal processing to convert 5.1-channel material for playback over a 7.1-channel loudspeaker array; and advanced video-switching circuits to ensure that the video quality isn't compromised by the AVR or controller. Ultra2-certified products also incorporate features such as Boundary Gain Compensation, a bass cut invoked when the loudspeakers must be placed near a wall.

Chapter 10

Multichannel Power Amplifiers

The power amplifier is the workhorse of your home-theater system. It takes in line-level signals from your controller's output and amplifies them to powerful signals that will drive your home-theater loudspeaker system. A power amplifier has line-level inputs to receive signals from the controller, and terminals for connecting loudspeaker cables. The power amplifier is the last component in the signal path before the loudspeakers.

A separate multichannel power amplifier can offer higher output power than the power amplifiers found in A/V receivers. The most powerful receivers can have outputs of perhaps 200 watts per channel; separate power amplifiers offer as much as 350Wpc.

A power amplifier is described by the number of amplifier channels in its chassis. The most common home-theater power amplifiers have five channels for powering the left, center, right, surround-left, and surround-right loudspeakers. The next most common number of channels is three. A 3-channel amplifier can power the front three loudspeakers, leaving a 2-channel amplifier to handle the two surround loudspeakers. If you already have a stereo (2-channel) amplifier, you can update your system for home theater by adding a 3-channel home-theater amplifier. Alternatively, a 3-channel amplifier can power the center and two surround channels, leaving your stereo amplifier to drive the left and right loudspeakers. As the Dolby Digital Plus and DTS-HD formats become more popular, power amplifiers with seven channels are becoming increasingly available. Note that if you upgrade your controller to a 7.1-channel unit, you can augment your 5-channel power amplifier with a stereo amplifier to power the surround-back speakers.

How to Choose a Home-Theater Power Amplifier

High-quality multichannel power amplifiers are big and heavy, and can be expensive. The more output power provided by the amplifier, the bigger, heavier, and more expensive it will be.

Most home-theater enthusiasts listen at moderate levels to loudspeakers of average sensitivity (85–88dB) in rooms of average size (3000–4500 cubic feet) with average furnishings (carpet, some drapes), and so will need a minimum of about 70W for each of the five, six, or seven channels.

Note that driving five, six, or even seven loudspeakers places extraordinary current demands on the amplifier. If the amplifier isn't up to the job of driving the loudspeakers adequately, the sound will be constrained on loud peaks and even distort. Just when the film soundtrack gets exciting, you hear the amplifier run out of power, distracting you from the movie. There's a way to reduce the strain placed on multichannel amplifiers: Use high-sensitivity loudspeakers. Remember that every 3dB increase in sensitivity is equivalent to doubling amplifier power.

More specific information on power amplifiers may be found in Chapter 8. The discussion of sound quality, build-quality, and matching to loudspeaker sensitivity applies equally well to multichannel amplifiers as it does to 2-channel amplifiers.

Loudspeakers for Home Theater

A home-theater loudspeaker system will perform double-duty, reproducing 2-channel music sources through the front left and right speakers, and film soundtracks through all five, six, or seven loudspeakers (plus an optional subwoofer). Home-theater loudspeakers can be evaluated using the criteria described in Chapters 4 and 9, with one difference: Film soundtracks present somewhat different challenges to a loudspeaker system than music. Specifically, film soundtracks are generally more dynamic than most music, and have much more low-frequency energy. Consequently, great home-theater sound requires loudspeakers that not only meet our criteria for good music reproduction, but also have a wide dynamic range and the ability to handle loud and complex signals without sounding smeared or confused.

The Center-Channel Speaker

Though the center-channel loudspeaker carries a large part of the film soundtrack—nearly all the dialogue, many effects, and some of the music—only recently have center-channel speakers been elevated from afterthoughts to recognition that they are the anchor of the entire home-theater loudspeaker system.

The center speaker is usually mounted horizontally on top of the video monitor (Fig.10-5). It can also be placed beneath the video monitor or mounted inside a wall above the video display, or, if you're using a front-projection system, behind an "acoustically transparent" screen.

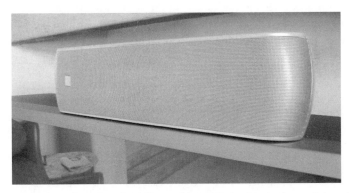

Fig.10-5 The center-channel speaker should be positioned flush with the front of the television set.

Because a stereo system uses only two loudspeakers—left and right—across the front, you may be wondering why you need a third loudspeaker between them for home theater. The center-channel loudspeaker provides many advantages in a home-theater system. First, it anchors dialogue and other sounds directly associated with action on the screen in the center of the sonic presentation. When we see characters speaking, we want the sound to appear to come from their visual images. Similarly, when sounds are panned (moved from one location to another) across the front, we want the sound to move seamlessly from one side to another. For example, if the image of a car travels from the left side of the screen to the right, the sounds of the car's engine and tires should travel with it, precisely tracking the car's movement. Without a center speaker, we may hear a gap in the middle as the car sounds jump from the left loudspeaker to the right. The center speaker makes sure on-screen sounds come from the screen.

A 2-channel stereo system is, however, capable of producing a sonic image directly between the two speakers. This so-called "phantom" center image is created by the brain when the same signal is present in both ears. A sound source directly in front of us in real life produces soundwaves that strike both ears simultaneously. The brain interprets these cues to determine that the sound source is directly in front of us. Similarly, two speakers reproducing the same sound send the same signal to both ears, fooling the brain into thinking the sound is directly in front of us.

For two loudspeakers to create this phantom center image, however, they must be precisely set up, and the listener must sit exactly the same distance from each of them—if you sit off to the right, the center image will pull to the right.

This problem is overcome by putting a center-channel speaker between the left and right speakers. Dialogue and onscreen sounds are firmly anchored on the screen for all listeners, not just those sitting in the middle. With three speakers across the front, someone sitting at the left end of the couch can still hear dialogue coming from the area of the screen—not just from the left loudspeaker. The center speaker also prevents the entire front soundfield from collapsing into the speaker closest to where you're sitting. Moreover, the center speaker provides a tangible sound source directly in front of you; your brain doesn't have to work to create a phantom image between the left and right speakers. Finally, the center speaker reduces the burden on the left and right loudspeakers. With three speakers reproducing sound, each can be driven at a lower level for cleaner sound.

Adding a Center Speaker to Your System

Although you can add a different brand of center speaker to your existing left and right speakers, you're better off with three matched speakers across the front of your home theater. Speakers all sound different from one another. No matter how good the quality of a speaker, it will have some coloration, or variations from accuracy, in its sound. Remember, you want to hear a seamless movement of sounds across the front soundfield. If the center speaker has a different sound from the left and right speakers, you'll never achieve a smooth and continuous soundfield across the front—when sound sources move from one side to another through the center speaker, the sound's character will abruptly change.

The solution is to buy three matched front-channel speakers. Their identical tonal characteristics will not only provide smooth panning of sounds, but also produce a more stable and coherent soundfield across the front of the room. If you already have high-quality left and right loudspeakers that you want to keep, buy a center-channel speaker made by the same manufacturer. They probably won't be as well-matched as a three-piece system, but the added center speaker is much more likely to sound similar to your existing left and right speakers. Separate center speakers sometimes use the same drivers (the raw speaker cones themselves) and other parts as the stereo speakers from the same manufacturer.

For state-of-the-art home theater, the three front loudspeakers should be identical. Although most home-theater enthusiasts will use a smaller center speaker that will fit on top of a television set and doesn't reproduce much bass, more ambitious systems use large, full-range center speakers. These speakers are hard to position, and are generally used only with front-projector systems; the center speaker can be placed behind a perforated projection screen, just as in a movie theater.

Left and Right Speakers

The left and right loudspeakers carry the majority of the film's musical score and many of the effects. And if you aren't using a subwoofer, nearly all the bass will be reproduced by the left and right speakers. Consequently, left and right speakers are often the largest speakers in a home-theater system.

Left and right speakers can also be small units that sit on speaker stands or on an entertainment cabinet. If that's the case, the small speakers must be used with a subwoofer to reproduce bass. The small left and right speakers reproduce the midrange and treble frequencies, and the subwoofer handles all the bass. Such a system, called a *subwoofer/satellite system* (Fig.10-6), is ideal if your available space is limited, or if you want the left and right loudspeakers to better blend into your decor. Satellite speakers can be small and unobtrusive, and the subwoofer can be tucked out of the way. If you

Fig.10-6 A subwoofer/satellite speaker system blends into home decor more easily than does a floor-standing, full-range system. (Courtesy Polk Audio)

build the three front speakers into a wall in a custom installation, they will most likely be satellites.

Surround Speakers

The surround loudspeakers are completely different in design and function from left, center, and right speakers. Their job is to re-create a diffuse atmosphere of sound effects to envelop us in a subtle sonic environment that puts us in the action happening on the screen. Unlike front and center speakers that anchor the sound onscreen, surround speakers should "disappear" into a diffuse "wash" of sound all around us.

A good example of how surround speakers create an atmosphere comes from the film *Round Midnight*. Toward the beginning of the movie the character François is outside a Paris jazz club in a driving rainstorm, listening to Dexter Gordon's character playing inside. The scene cuts between the intimate sound of the jazz club and the rainy Paris street. Inside the club, the sound is direct and immediate to reflect the camera's perspective of just a few feet from the musical group. When the scene cuts to the street, we are surrounded by the expansive sound of rain, cars driving by, people talking as they walk past—in other words, all the ambiance and atmosphere of a rainy night in Paris. This envelopment is largely created by the surround speakers. We don't want to hear the rain and street sounds coming from two locations behind us, but to be surrounded by the sounds, as we'd hear them in real life. The surround speakers perform this subtle, yet vital role in home theater.

Surround speakers envelop us by their design and placement in the home-theater room. They are best located to the side or rear of the listening position, and several feet above ear level. Because they don't have to reproduce bass, surround speakers can be small and unobtrusive, and are often mounted on or inside a wall.

Dipolar and Bipolar Surround Speakers

Most of the surround speakers' ability to wrap us in sound comes from their *dipolar* design. Dipolar simply means that they produce sound to the front and rear equally. While front speakers have one set of drivers that project sound forward, dipolar surround speakers have two sets of drivers, mounted front and back. This arrangement produces a directional pattern that fires to the front and back of the room (Fig.10-7). Because the surround speakers are positioned to the sides and fire to the front and back of the room, we hear no direct sound from them. The listener sits in the surround speakers' "null"—the point where they don't directly project sound. Instead of hearing direct sound from the surround speakers, we hear their sound after it's been reflected, or bounced off, the room's walls and furnishings. The surround speakers' dipolar directional pattern makes their sound diffuse and harder to localize. You shouldn't be able to tell where a properly set-up surround speaker is just by listening.

You may see surround speakers called either *dipolar* or *bipolar*. Although both produce sound equally from the front and rear, the bipolar speaker produces sound

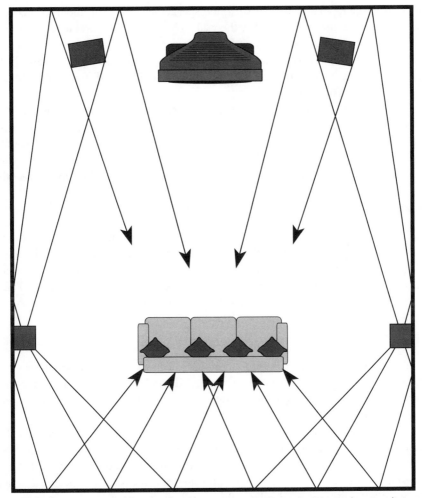

Fig.10-7 The surround speakers' dipolar radiation pattern results in the listener hearing their sounds only after being reflected from a room surface.

from the rear that is in-phase with the front sound. A dipolar produces sound from the rear that is out-of-phase with the front sound. In-phase means that the front and rear waves have the same polarity: When the front-firing woofer moves forward to create a positive pressure wave, the rear-firing woofer also moves forward. The opposite is true in a dipole: When the front-firing woofer moves forward, the rear-firing woofer moves backward. In a dipole, the front and back waves are identical, but have the opposite polarity. Both bipolar and dipolar designs achieve the objective of enveloping the viewer/listener in the film's aural ambiance, but the dipolar type is preferred because it creates a greater "null," or reduction in sound, at the side of the speaker that faces the listener.

A true dipolar or bipolar surround speaker can't be flush-mounted inside a wall. Instead, it must be mounted *on* the wall so it can radiate sound toward the front and rear of the room. Some surround speakers approximate a dipolar radiation pattern by mounting the drivers on an angled baffle. Although not true dipoles, these flush-mounted surround speakers accomplish the goal of creating a diffuse ambiance *and* fit unobtrusively in your living room.

The amount of sound produced by the surround speakers is adjustable through the receiver's or controller's remote control. Their volume level should be set so that you never hear them directly, but if you turned them off, the soundfield would collapse toward the front. All too often, consumers set surround speaker levels too high; their thinking seems to be that if they *bought* speakers, they should be able to *hear* them.

Surround-Back Speakers for 7.1-Channel Sources

As described earlier, the Dolby Digital EX and DTS-ES formats add one or two additional surround speakers behind the listener. Dolby Digital Plus and DTS-HD are full 7.1-channel formats that require two additional surround-back speakers. THX recommends that two speakers be used, and that their radiation patterns be dipolar. The surround-back speakers are usually identical to the left- and right-surround speakers.

Subwoofers

A subwoofer is a speaker that reproduces only low bass. Subwoofers are usually squarish or rectangular cabinets that can be positioned nearly anywhere in the home-theater room. A good subwoofer adds quite a bit of impact and visceral thrill to the home-theater experience.

Most subwoofers for home-theater use are *powered* or *active* subwoofers. These terms describe a subwoofer with a built-in power amplifier that drives its large woofer cone. The powered subwoofer takes a line-level output from the jack marked SUB-WOOFER OUT on your A/V receiver or controller, and converts that electrical signal into sound. Because powered subwoofers have built-in amplifiers, they must be plugged into an AC wall outlet. (See Chapter 9 for more on subwoofers.)

Setting up a Home Theater

How a home-theater system is set up and calibrated has an enormous influence on the quality of the sound you hear. Just as with a 2-channel music system, ideal loudspeaker placement and careful attention to detail will reward you with superior performance. The acoustic principles described in the next section and in Chapter 12 apply equally well to a multichannel audio system.

Basic Setup

The first step in setting up your system is deciding where to install it. More than likely, the room where you watch television will become the home-theater room. But this room may have been arranged for casual TV watching, not for experiencing the full impact of what a home-theater system can deliver. Now is the time to rethink the room's layout and consider making the home-theater system the room's central focus.

The home-theater room should be arranged with the video monitor and front loudspeakers at one end, the couch or viewing chairs at the other. Ideally, the couch will be positioned partway into the room and not against the back wall. This location is better acoustically (excessive bass build-up occurs near the back wall), and allows the surround speakers to better do their job of enveloping you in the film's soundtrack. Of course, the setup should provide room for the left and right speakers on either side of the video monitor.

The optimum viewing distance between you and the video monitor is determined by the video monitor's size. The smaller the monitor, the closer you should sit to maintain the maximum image area in your field of view. Sitting too close to a large monitor, particularly a rear-projection set, exposes the pixel structure inherent in a video presentation. If you are watching high-definition programming on an HD set, you can sit much closer without seeing the pixel structure. The best viewing distance offers the largest possible picture without making that picture look grainy. A good rule of thumb is to sit three-to-four picture *heights* away from the screen. You should also try to sit no more than 15° off to the side of the video monitor. Plasma panels, with their wider field of view, maintain a good picture even when they are watched from an extreme angle.

Once you've decided on where the couch and video monitor will be positioned, you'll need to determine the best place for your equipment. A simple home-theater system will have a VCR or hard-disk digital video recorder, A/V receiver, satellite receiver, and DVD player. These components can be easily accommodated in a TV stand or rack.

Acoustical Treatment

Before we get to the specifics of loudspeaker placement, here are a few guidelines on how to make your home-theater room sound better. First, if you have a bare tile or wood floor between you and the front three speakers, covering the floor with a rug is probably the single most important thing you can do to improve your room's acoustics. The rug absorbs unwanted acoustic reflections from the floor, which improves dialogue intelligibility and adds to a sense of clarity. Similarly, absorbing the acoustic reflections from the side walls between you and the left and right speakers also helps you to get better sound. Bookcases and hanging rugs are both effective in absorbing or diffusing side-wall reflections. Generally, the front of the room should be acoustically absorptive, with drapes, carpets, and hanging rugs.

Conversely, the back of the room should be more acoustically reflective. Bare walls behind the listening/viewing position tend to reflect sound and add to the feeling of spaciousness generated by the surround speakers. The optimum condition is when the wall behind the listening position diffuses the sound. Bookcases and media-storage racks make great diffusers. The sounds produced by a movie theater's array of surround speakers all reach your ears at different times, increasing the feeling of envelopment. But because a home-theater system uses just two surround speakers, we must help them achieve their goal by providing reflective surfaces near them.

Speaker Placement

Where speakers are located in your room has a huge influence on the quality of sound you will achieve. Correct speaker placement can make the difference between good and spectacular performance from the same equipment. If you follow a few simple speaker-placement techniques, you'll be well on your way toward getting the best sound from your system. Correct speaker placement is the single most important factor in achieving good sound.

Surround-Speaker Placement

Let's start with the most difficult of a home-theater system's six speakers: the two surround speakers. As I said earlier, the surround speakers are usually dipolar. That is, they produce sound equally to the front and the rear of the speaker cabinet. Dipoles produce a "null," or area of no sound, at the cabinet sides. Here's a simple rule for positioning dipolar surround speakers no matter what the installation: Point the surround speakers' null at the listening position. Never point the surround speakers directly at the listening position.

Ideally, the surround speakers would be located on the side walls directly to the side of the listening position and at least two feet above the listeners' heads when seated. Dipolar surround speakers should never be located in front of the listening position, where they could be heard directly; the surround speakers' sound should be reflected around the room before reaching the listener. This helps to mimic the sound of a movie theater's surround array with just two speakers, and to create the envelopment that's so important to the home-theater experience.

Keep in mind that reflections of signals from the front speakers are bad; reflections of signals from the rear speakers are good. Well-positioned surround speakers make a tremendous difference in the quality of the home-theater experience by re-creating the sense of ambience and space we hear in a movie theater.

Surround-Back Speaker Placement for 7.1-Channel Sources

According to Lucasfilm, the surround channels are best reproduced by four identical dipolar speakers. The two surround speakers remain at their normal position (normal for 5.1-channel reproduction), which is 110°, or slightly behind the listener on the side walls. The surround back speakers are located at 150°.

The two surround-back speakers should be arranged so that their facing sides are in-phase with each other. That is, the side of the dipole that would face the back of the room if mounted on a sidewall should face toward the room's centerline when the speaker is mounted on the rear wall. This arrangement keeps the drivers of the two surround back speakers that face each other in-phase, contributing to their ability to create an image between them directly behind the listening position. If you're using a single surround-back speaker, position it directly behind the listener at 180°.

Center-Speaker Placement

Correct positioning of the center speaker results in better dialogue intelligibility, smoother movement of onscreen sounds, and a more spacious soundstage. As just mentioned, reflections from the front speakers that reach the listener work against good sound. This is particularly true of the center speaker, which carries the all-important dialogue. Because most center-channel speakers are mounted on top of a direct-view TV or rear projector—both of which are highly reflective—there's the potential for unwanted reflections from the video monitor.

To reduce reflections from the center speaker, align its front edge absolutely flush with the front of the video monitor so that they form as smooth a surface as possible. This positioning reduces a phenomenon called *diffraction,* which is the re-radiation of sound when the soundwaves encounter a discontinuity. (Specifically, diffraction introduces variations in the speaker's frequency response.) Think of waves rippling in a pond: If you put a stick in the water, the waves will strike the stick and be re-radiated around it. The same thing happens when the sound from the center speaker encounters the "stick" of the video monitor: the monitor re-radiates that sound toward the listener, making the overall sound less accurate. That's why making the center speaker flush with the front of the video monitor is so important (Fig.10-5 earlier).

High-end loudspeaker manufacturers carefully design their enclosures to reduce diffraction. You'll notice that high-quality loudspeakers have smooth surfaces around the speaker cones. Some have rounded fronts and even the mounting bolts securing the individual drivers within the speaker are recessed to reduce diffraction from the speaker cabinet. The loudspeaker on this book's cover is an extreme example of an enclosure designed to minimize diffraction.

You should also tilt the center speaker down toward the listening position if it is mounted on a large rear-projection set, or on a direct-view TV positioned on a tall stand. If the center speaker is located below the video monitor, point it up toward the listeners. Direct the center speaker's sound toward the height of the listeners' ears when seated. Finally, be sure the center speaker is less than 2' different in height from the left and right speakers, measured from each speaker's tweeters.

Left and Right Speaker Placement

The left and right speakers should be pulled slightly in front of the video monitor, and not be in a straight line with it (as some owner's manuals erroneously advise). Positioning the left and right speakers so that they form a gentle arc with the center speaker has two advantages. First, acoustic reflections from the video monitor are reduced. Second, the left, right, and center speakers are now all the same distance from the listener. The sounds from each of the three front speakers will reach the listener at the same time.

The speaker-placement guidelines for 2-channel stereo presented in Chapter 12 apply equally well to the left and right speakers in a multichannel system.

Subwoofer Placement and Connection

Where the subwoofer is positioned in your room will greatly affect the bass smoothness, clarity, and impact. An ideal—but often impractical—position is right next to the listening seat. This placement results in the listener hearing more of the direct wave-launch from the woofer and less of the woofer's sound after it has been modified by the room (this phenomenon is described in detail in Chapter 12).

A rule of thumb says that for the smoothest bass response, don't position a subwoofer at the same distance from two walls. For example, if you want to put the sub on the wall behind your video display, don't put the sub mid-way between the two walls. The guidelines presented in Chapter 12 for achieving the smoothest bass from left and right loudspeakers applies equally well to subwoofer positioning.

Many subwoofers offer several ways of connecting the sub to the rest of your system. The first connection method uses the speaker-level output from your receiver. You connect speaker cables between your AVR and the sub, and between the sub and your left and right speakers. The subwoofer's internal crossover splits up the frequency spectrum, filtering bass from the signal driving your left and right speakers. In this connection method, the crossover that divides the frequency spectrum operates on the high-level signal (one powerful enough to drive speakers).

The second, and preferable, method is to connect the AVR's SUBWOOFER OUT-PUT jack to the subwoofer's line-level input. The AVR's internal crossover (bass management) splits the frequency spectrum, sending bass to the subwoofer at line-level through the SUBWOOFER OUTPUT jack, and powering the left and right speakers from the AVR's internal amplifiers. Note that this method requires that the subwoofer be of the active type, with an integral power amplifier to drive the woofer cone.

This second method avoids subjecting the signal driving the left and right speakers to filtering by the subwoofer. The result is cleaner sound.

A third subwoofer-connection method, used with a separate controller and power amplifier, puts the subwoofer in the line-level signal path between the controller and the amplifier. The controller's left and right line-level outputs feed the subwoofer's line inputs; the sub's line outputs drive the left and right power amplifiers, but with bass removed from the signal by the subwoofer's crossover.

Some subwoofers have separate PROCESSOR and LINE inputs. The PROCESSOR input (a single jack) accepts the subwoofer output from an AVR or controller; the LINE input is stereo, and is connected to a second output from your preamplifier. When watching movies, set the subwoofer's input switch to PROCESSOR. When listening to music, set the sub's input switch to LINE. This assumes, of course, that you prefer listening to music with the subwoofer engaged.

Calibrating a Home Theater

Once your system is connected, you'll need to configure it for your particular room and speaker system. This is done through a series of menus or graphical icons generated by the A/V receiver or A/V controller and appearing on your video monitor. You use the receiver's remote control to make selections from the onscreen display. Every A/V receiver or controller includes instructions for setting up your system; the following section explains why you make certain selections, and in what circumstances.

When you turn on your system for the first time, don't be alarmed if you don't hear any sound when you expect to. (This usually happens when I first set up a product for review.) Don't turn up the volume to a high level and then start pushing buttons—when you push the right button, the sound will blast you out, possibly damaging your speakers, your amplifier or receiver, and/or your hearing. Turn the volume to a low level and make sure the correct source is selected. Check to make sure the MUTE and TAPE MONITOR buttons *aren't* pressed. The lack of sound is probably due to one of these three incorrect settings.

Bass Management

The first calibration step is configuring the receiver's *bass management* settings for your particular loudspeaker system. Bass management describes a receiver function that diverts bass from small loudspeakers to the subwoofer, if the system uses one. The first step in system setup is telling the receiver what kind of loudspeakers you have so it can direct bass appropriately. Most of today's receivers ask you whether each of the speakers is LARGE or SMALL (see Fig.10-3 earlier). By selecting SMALL for the left and right speakers, you're telling the receiver to keep low-bass signals out of those speakers. This configuration is used when you have a subwoofer connected to the system to reproduce low bass instead of the left and right speakers. You'll also need to answer YES when asked if the system includes a subwoofer. The bass-management option labeled "THX" in THX-certified products automatically sets all speakers to SMALL and engages the subwoofer.

If your system includes full-range left and right speakers and no subwoofer, answer LARGE in the setup menu when asked if the left and right speakers are LARGE or SMALL. The LARGE setting directs bass to the left and right speakers.

The setup menu will also ask if the surround speakers are LARGE or SMALL. Nearly every installation will use the SMALL setting. Only if you have full-range, floor-standing speakers should you answer LARGE.

Although full-range left and right speakers deliver adequate bass extension and dynamics, you won't hear the full bottom-end impact of today's soundtracks without a subwoofer to reproduce the LFE channel (the LFE channel is the ".1" channel in 5.1- or 7.1-channel sound). That channel contains high-impact, low-bass sounds that can only be correctly reproduced by a dedicated subwoofer.

Chapter 10

Setting Individual Channel Levels

Next you'll need to individually set the loudness of each of the five channels (six if you're using a subwoofer), or seven channels (eight if you're using a subwoofer) in a system equipped with Dolby Digital Plus and DTS-HD decoding. In addition to providing an overall volume control, all A/V receivers provide adjustment of the individual channel volumes. This process begins by turning on the "test signal," a noise-like sound generated by the A/V receiver. (Dolby licensing requires that the receiver or controller incorporate this noise generator.) The noise is produced by each speaker in turn. Ideally, the noise signal's volume should be the same for each speaker when you're sitting at the listening/viewing position. If it isn't, you can adjust the volume of each speaker independently using the receiver's remote control. Individually adjusting the channels lets you compensate for different loudspeaker sensitivities, listening-room acoustics, and loudspeaker placements.

Although setting the individual-channel levels by ear will get your system in the ballpark, a more precise calibration can be achieved by using a Sound Pressure Level (SPL) meter. Available from RadioShack for about $30, an SPL meter lets you accurately calibrate your home-theater system. RadioShack offers digital and analog SPL meters; buy the easier-to-read analog type (catalog #33-2050). Switch the meter to the C-WEIGHTED position, SLOW RESPONSE, and set the knob to 70. Turn on the test noise on your A/V receiver and set each channel's volume until the meter's display reads 5. This indicates that the noise is being reproduced at a level of 75dB. Repeat this procedure for each channel.

A powered subwoofer usually has a knob on its back panel for setting loudness. You should set this volume knob in a position that provides roughly the correct loudness so that you can fine-tune the volume through the receiver's channel-level control. When you set the subwoofer's level, the meter will be hard to read because the indicator will be jumping around (it's reading a low frequency). Stare at the meter for a few minutes as it jumps to get an idea of where the average level is.

These settings will get you very close to the optimum volume for each channel, but you should use your ear and some well-recorded film soundtracks in making the final adjustment. If you find that dialogue is hard to hear, a 2dB boost in the center-channel level will help bring it out (although the soundstage will be focused more in the center than spread out across the front of the room). If the bass is thumpy and boomy, turn down the subwoofer. Don't be afraid to make small adjustments to the volume of the subwoofer, center channel, or surround speakers.

Although your ears should be the final judge when setting the individual channel levels, calibrating your system first with the SPL meter (or by ear with the test noise) will at least get you started from the right place. The most common mistakes in setting channel levels are a subwoofer level set too high and too much volume from the surround speakers. The bass shouldn't dominate the overall sound, but instead serve as the foundation for music and effects. Many listeners think the more bass you hear, the more "impressive" the sound. In reality, a constantly droning bass is fatiguing, and robs the soundtrack of impact and surprise when the filmmakers *want* you to hear bass. Low frequencies are used as punctuation in a film soundtrack; by keeping the subwoofer level appropriate, you'll achieve a more accurate and satisfying sound than if the subwoofer is constantly droning away. Though it's understandable that you paid for this big box in your living room and you want to hear what it does, try to avoid the temptation to set the subwoofer level too high.

Similarly, listeners who have never had surround speakers in their homes think they should be aware of the surround speakers at all times. In truth, film soundtracks don't always contain signals in the surround channels; long stretches of the movie may have *nothing* in the surrounds. But even when a signal is driving the surround speakers, you should barely notice their presence. Remember, the surround channels provide a subtle ambience and envelopment. If you're consciously aware of the surround speakers because they're set too loud, the illusion they're supposed to be creating is diminished. Just as it's a temptation to set a subwoofer's level too high, don't turn up the surround speakers to the point where you hear them. Note that you won't hear *any* sound from the surround speakers unless the source program has been Dolby- or DTS-encoded (unless you have Pro Logic or other processing engaged).

Adding Home Theater without Compromising Music Performance

As Editor-in-Chief of *The Absolute Sound* and *The Perfect Vision* magazines (the latter for six years until late 2006), my job involves evaluating cutting-edge, state-of-the-art 2-channel audio products as well as high-performance home-theater components. Clearly, the performance of my 2-channel system cannot be compromised by the presence of home-theater products. Here's what I've done to integrate home theater into my music-playback system.

First, my video display is a front-projection system with a retractable, motorized screen. The projector is at the back of the room, and when I listen to music, the screen is rolled up into a small enclosure. With the screen in the lowered position, soundstage depth is compromised, as is the precision of image placement. If you use a front projector, a motorized screen's ability to retract for music listening is a big benefit. If you must use a fixed screen, drapes that can be drawn across the screen when you play music are effective at preventing the screen from reflecting sound.

Those who use a flat-panel television or rear-projector big-screen are faced with the challenge of having a large, acoustically reflective object near the loudspeakers. As described in Chapter 12, absorbing acoustic reflections at the loudspeaker-end of the room is important. To minimize the television's degradation of the sound, move the television back as far as possible, and bring the left and right loudspeakers forward.

As mentioned earlier, an analog-bypass mode on a controller is absolutely essential if you want to listen to analog sources with the highest possible fidelity. This feature passes analog input signals through the controller without converting the signal to digital and then back to analog. Keep in mind, however, that even the very best multichannel controllers fall short of the performance standards set by high-end 2-channel preamplifiers, even in analog-bypass mode. If you want uncompromised musical performance, you'll need a controller *and* a separate 2-channel preamplifier. The preamplifier should have a "theater pass-through" mode that sets one of the inputs at unity gain (the output signal's amplitude is the same as the input signal). The left and right outputs from your controller drive this unity-gain input, and the preamplifier is connected to the power amplifier in the usual way. When watching movies, it's as though your preamp isn't there. When listening to music, it's as though your controller isn't there. With this technique, shown in Fig.2-4 on page 12, your analog sources never go through the controller. If you choose this option, you don't need a controller with analog bypass, except on the multichannel input.

Addendum: Surround-Sound Formats

I've broken out a full explanation of the various surround-sound formats for those readers interested in more technical detail. Even if you're not technically inclined, this section includes useful information about choosing among the vast array of decoding formats in today's AVRs and controllers.

We've already covered Dolby Pro Logic, Dolby Digital, and DTS in the body of this chapter. Now let's look at variations on those formats.

In mid-2001, Dolby Laboratories made available a more sophisticated version of Pro Logic decoding, called Pro Logic II. The idea behind Pro Logic II was to create from 2-channel sources a listening experience similar to that of a discrete 5.1-channel digital format. And with more consumers having 5.1-channel playback available to them for reproducing music sources, Pro Logic II attempts to deliver multichannel sound from 2-channel recordings, even those that have not been surround-encoded.

Found on most A/V receivers and controllers made after early 2002, Pro Logic II offers improved performance over its predecessor in several areas. First, Pro Logic II delivers full-bandwidth stereo surround channels rather than the bandwidth-limited monaural surround channel of conventional Pro Logic decoding. This attribute provides a more enveloping soundfield, greater precision in the placement and pans (movements) of sounds behind the listener, and more natural timbre of sounds reproduced by the surround channels. In this respect, Pro Logic II emulates the experience of listening to a 5.1-channel discrete digital source.

Pro Logic II also uses more sophisticated "steering" circuits that monitor the level in each channel and selectively apply attenuation (reduction in level) to prevent sounds in one channel from leaking into another channel. Pro Logic IIx, announced in late 2003, creates 7.1-channel playback from 2-channel and 5.1-channel sources.

Not to be outdone, DTS has developed a suite of surround-decoding formats that either enhance the DTS experience or provide decoding of non-DTS sources such as conventional stereo.

DTS Neo:6 Music and *Neo:6 Cinema* are decoding algorithms that convert stereo or Dolby Surround encoded 2-channel sources into multichannel surround sound. Neo:6 Music leaves the front left and right channel signals unprocessed for the purest reproduction, and extracts center and surround-channel signals from the 2-channel source. DTS recommends this mode for all 2-channel sources, such as CD and FM broadcasts.

Neo:6 Cinema is similar to Dolby Pro Logic II decoding, and can be used with Dolby Surround-encoded sources. Neo:6 Cinema has a much larger effect on the signal, rearranging the signal distribution among the front three channels. Both Neo:6 Music and Neo:6 Cinema create a 7.1-channel signal from 2-channel sources.

These decoding algorithms are very useful to those, like me, who greatly enjoy musical performances on DVD. When a DVD gives me the choice of listening to the Dolby Digital or 2-channel Surround-encoded linear pulse-code modulation (LPCM) mix, I always opt for the 2-channel mix. Although I lose all the advantages of Dolby Digital (complete channel separation, for example), I hear smoother treble, less grainy instrumental and vocal textures, and a greater sense of space. That's because Dolby Digital uses 384,000 bits per second (or 448,000 bits per second) to encode all 5.1 channels; the linear PCM track uses 1.536 million bits per second to encode just two channels. When using Pro Logic II and DTS Neo:6 to decode these two channels, however, I still hear excellent spatial resolution, but with the more natural timbres made possible by the linear PCM track's much higher bit rate.

In 2004, Dolby Labs and Lucasfilm jointly developed the *Dolby Digital EX* format. Dolby Digital EX encodes a third surround channel in the existing left and right surround channels in the Dolby Digital signal. This additional surround channel, called surround back, is decoded on playback, and the signal drives a loudspeaker (or two loudspeakers) located directly behind the listener. Dolby Digital EX allows film-makers to more precisely position and pan (move) sounds around the room. For example, the sound of an object moving directly overhead from the front of the room to the back tends to become smeared; the sounds starts in the front speakers, then splits along the sidewalls as it moves toward the rear of the room. By adding a third surround channel directly behind the listener, these "flyovers" can be made more realistic. Note that an EX-encoded soundtrack is compatible with all Dolby Digital playback equipment because the addition of the surround-back channel is encoded into a conventional 5.1-channel Dolby Digital signal. Dolby Digital EX is not a 6.1-channel format; it is still 5.1-channels, but with the additional channel encoded within the 5.1-channel datastream.

DTS' equivalent format is called *DTS-ES*. You'll also see it called *DTS-ES Matrix*, because the additional surround channel is matrix-encoded into the existing left and right surround channels. DTS also offers *DTS-ES Discrete*, in which the surround-back channel is a discrete channel in the ES-Discrete bitstream. DTS-ES Discrete is a true 6.1-channel format. The advantage of a discrete format is complete separation between channels. Matrix surround systems have limited channel separation.

Note that Dolby Digital and DTS are "lossy" formats, meaning that some information is lost in the encoding and decoding process. This loss of information is intentional so that the number of bits required to represent the signal can be dramatically reduced. Consequently, Dolby Digital and DTS have reduced fidelity to the source.

Chapter 10

The massive storage capacity of HD DVD and Blu-ray Disc have allowed Dolby Laboratories and DTS to develop better-sounding surround-sound formats with higher bit rates than Dolby Digital and DTS. From Dolby Laboratories, we have two new formats called *Dolby Digital Plus* and *Dolby TrueHD*. Dolby Digital Plus is an extension of the conventional Dolby Digital we've been listening to for years on DVD and HD television shows. Dolby Digital Plus offers more channels and a higher bit rate for better sound quality. Where Dolby Digital was limited to a maximum bit rate of 640kbps (kilobits per second), Dolby Digital Plus allows scalable bit rates up to 3Mbps (million bits per second) on HD DVD and up to 6Mbps on Blu-ray. In addition to more bits per second, Dolby improved the encoding algorithms in Dolby Digital Plus for better sound.

For the ultimate in sound quality, Dolby has introduced Dolby TrueHD, which delivers high-resolution multichannel audio with perfect bit-for-bit accuracy to the source. With TrueHD, you will hear in your home sound quality identical to what the engineers heard in the studio. TrueHD decoding is an option, rather than a requirement, in HD DVD and Blu-ray players. The availability of a consumer release format that delivers uncompressed high-resolution multichannel audio is nothing short of a revolution in home entertainment.

With more than 40 million Dolby Digital decoders in use throughout the world, Dolby made sure the new formats were backward-compatible with your receiver or controller. The new players output a conventional Dolby Digital or DTS bitstream on the familiar coaxial or optical digital outputs for connection to your receiver. This output will likely sound a little better than Dolby Digital from DVD because it always operates at the highest possible data rate of 640kbps.

DTS has introduced its own high-resolution surround-sound audio format for HD DVD and Blu-ray called DTS-HD. This new format is an extension of conventional DTS, offering scalable bit rates from conventional DTS at 754kbps, all the way to 3Mbps on HD DVD and 6Mbps on Blu-ray. A variant of DTS-HD, called DTS-HD Master Audio, can operate at higher bit rates (up to 18Mbps on HD DVD and 24Mbps on Blu-ray) for true lossless encoding of high-resolution multichannel audio. The "Master Audio" tag signifies lossless encoding. As with Dolby TrueHD, HD DVD players are not required to implement DTS-HD Master Audio; however, if a player encounters that format on a disc, it must at least deliver the DTS core audio stream (that is, good old standard DTS).

All HD DVD and Blu-ray Disc players are required to have Dolby Digital Plus and DTS-HD decoding. Unlike DVD, however, that required a Dolby Digital track on the disc, the new disc formats don't require any specific format. That decision is left to the disc's producers.

11

Cables, Racks, and AC Conditioners

In addition to the main product categories described in previous chapters—source components, preamplifiers, power amplifiers, and loudspeakers—you'll need three additional components to complete your audio system—cables and interconnects, an equipment rack, and an AC power conditioner. I refer to these products as "components" because they are important contributors to getting the best sound from your system and should be thought of as full-fledged components rather than as optional "accessories."

Cables and Interconnects

Loudspeaker cables and line-level interconnects are an important but often overlooked link in the music playback chain. The right choice of loudspeaker cables and interconnects can bring out the best performance from your system. Conversely, poor cables and interconnects—or those not suited to your system—will never let your system achieve its full musical potential. Knowing how to buy cables will provide the best possible performance at the least cost.

Let's start with an overview of cable and interconnect terms.

Cable: Often used to describe any wire in an audio system, "cable" more properly refers to the conductors between a power amplifier and a loudspeaker. Loudspeaker cables carry a high-current signal from the power amplifier to the loudspeaker.

Interconnect: Interconnects are the conductors that connect line-level signals in an audio system. The connection between source components (turntable, CD player, tuner) and the preamplifier, and between the preamplifier and power amplifier, are made by interconnects.

Unbalanced Interconnect: An unbalanced interconnect has two conductors and is usually terminated with RCA plugs. Also called a *single-ended* interconnect.

Balanced Interconnect: A balanced interconnect has three conductors instead of two, and is terminated with 3-pin *XLR* connectors. Balanced interconnects are used only between components having balanced inputs and outputs.

Digital Interconnect: A single interconnect that carries a stereo digital audio signal, usually from a CD transport or other digital source to a digital processor. A digital interconnect can also carry multichannel surround-sound, such as from a DVD player to an A/V receiver.

Bi-wiring: Bi-wiring is a method of connecting a power amplifier to a loudspeaker with two runs of cable instead of one.

RCA Plug and Jack: RCA plugs and jacks are the most common connection termination for unbalanced signals. Virtually all audio equipment has RCA jacks to accept the RCA plugs on unbalanced interconnects. RCA jacks are mounted on the audio component's chassis; RCA plugs are the termination of unbalanced interconnects.

XLR Plug and Jack: XLR plugs are three-pin connectors terminating a balanced interconnect. XLR jacks are chassis-mounted connectors that accept XLR plugs.

Binding Post: Binding posts are terminations on power amplifiers and loudspeakers that provide connection points for loudspeaker cables.

Spade Lug: A flat, pronged termination for loudspeaker cables. Spade lugs fit around power-amplifier and loudspeaker binding posts. The most popular kind of loudspeaker cable termination.

Banana Plug and Jack: Banana plugs are sometimes found on loudspeaker cables in place of spade lugs. Banana plugs will fit into five-way binding posts or banana jacks. Many European products use banana jacks on power amplifiers for loudspeaker connection.

How to Choose Cables and Interconnects

Ideally, every component in the system—including cables and interconnects—should be absolutely neutral and impose no sonic signature on the music. As this is never the case, we are forced to select cables and interconnects with colorations that counteract the rest of the system's colorations.

For example, if your system is a little on the bright and analytical side, mellow-sounding interconnects and cables can take the edge off the treble and let you enjoy the music more. If the bass is overpowering and heavy, lean- and tight-sounding interconnects and cables can firm up and lean out the bass. A system lacking palpability and presence in the midrange can benefit from a forward-sounding cable.

Selecting cables and interconnects for their musical compatibility should be viewed as the final touch to your system. A furniture maker who has been using saws, planers, and rasps will finish his work with steel wool or very fine sandpaper. Treat cables and interconnects the same way—as the last tweak to nudge your system in the right direction, not as a Band-Aid for poorly chosen components.

Cables and interconnects won't correct fundamental musical or electrical incompatibilities. For example, if you have a high-output-impedance power amplifier driving current-hungry loudspeakers, the bass will probably be soft and the dynamics constricted. Loudspeaker cables won't fix this problem. You might be able to somewhat correct the soft bass with the right cable, but it's far better to fix the problem at the source—a better amplifier/loudspeaker match.

Good cables merely allow the system's components to perform at their highest level; they won't make a poor system or bad component match sound good. Start with a high-quality, well-chosen system and select cables and interconnects that allow your system to achieve its highest musical performance. Remember, a cable or interconnect can't actually effect an absolute improvement in the sound; the good ones merely do less harm.

A typical hi-fi system will need one pair of loudspeaker cables (two pairs for bi-wiring), one long pair of interconnects between the preamplifier and power amplifier, and several short interconnect pairs for connections between source components (such as a turntable or CD player) and the preamplifier. Systems based on an integrated amplifier obviously don't need the long interconnect between a preamplifier and power amplifier.

Once you've got a feel for how your system is—or will be—configured, make a list of the interconnects and cables you'll need, and their lengths. Keep all lengths as short as possible, but allow some flexibility for moving loudspeakers, putting your preamp in a different space in the rack, or other possible changes. Although you'll want to keep the cables and interconnects short for the best sound, there's nothing worse than having interconnects 6" too short. After you've found the minimum length, add half a meter for flexibility.

Interconnects are often made in standard lengths of 1, 1.5, and 2 meters. These are long enough for source-to-preamplifier connections, but too short for many preamplifier-to-power-amplifier runs. These long runs are usually custom-made to a specific length. Similarly, loudspeaker cables are typically supplied in 8' or 10' pairs, but custom lengths are readily available. It's best to have the cable manufacturer terminate the cables (put spade lugs or banana plugs on loudspeaker cables, and RCA or XLR plugs on interconnects) rather than trying to do it yourself.

Concentrate your cable budget on the cables that matter most. The priority should be given to the sources you listen to most. For example, you may not care as much about the sound of your tuner as you do your CD player. Consequently, you should spend more on interconnects between the CD player and preamplifier than between the tuner and preamp. And because all your sources are connected to the power amplifier through the interconnect between the preamplifier and power amplifier, this link must be given a high priority. But any component—even a tuner—will benefit from good interconnects.

Most dealers will let you take home several cables at once to try in your system. Take advantage of these offers. Some mail-order companies will send you many cables to try: you keep the ones you want to buy—if any—and return the others. Compare inexpensive cables with expensive ones; sometimes manufacturers have superb cables that sell for a fraction of the price of their top-of-the-line products.

If you're starting a system from scratch, selecting cables is more difficult than replacing one length in your system. Because different combinations of cables will produce different results, the possibilities are greatly increased. Moreover, you don't have a baseline reference against which to judge how good or bad a cable is. In this situation, the best way of getting the ideal cables for your system is your dealer's advice. Try the cables and interconnects he suggests, along with two other brands or models for comparison.

Chapter 11

How Much Should You Spend on Cables and Interconnects?

At the top end of the scale, cable and interconnect pricing bears little relationship to the cost of designing and manufacturing the product. Unlike other audio products, whose retail prices are largely determined by the parts cost (the retail price is typically a set multiple of raw parts' cost), cables and interconnects are sometimes priced according to what the market will bear. This trend began when one company set its prices vastly higher than everyone else's—and saw its sales skyrocket as a result. Other manufacturers then raised *their* prices so they wouldn't be perceived as being of lower quality. Although some very expensive cables and interconnects are worth the money, many cables are ridiculously overpriced.

The budget-conscious audiophile can, however, take advantage of this phenomenon. Very often, a cable manufacturer's lower-priced products are very nearly as good as its most expensive models. The company prices their top-line products to foster the impression of being "high-end," yet relies on its lower-priced models for the bulk of its sales. When shopping for loudspeaker cables and interconnects, listen to a manufacturer's lower line in your system—even if you have a large cable budget. You may be pleasantly surprised.

Because every system is different, it's impossible to be specific about what percentage of your overall system investment you should spend on cables and interconnects. Spending 5% of your system's cost on cables and interconnects would be an absolute minimum, with about 15% a maximum figure. If you choose the right cables and interconnects, they can be an excellent value. But poor cables on good components will give you poor sound and are false economy.

Again, I must stress that high cost doesn't guarantee that the cable is good or that it will work well in your system. Don't automatically assume that an expensive cable is better than a low-priced one. Listen to a wide variety of price levels and brands. Your efforts will often be rewarded with exactly the right cable for your system at a reasonable price.

What to Listen For

Cables must be evaluated in the playback system in which they will be used. Not only is the sound of a cable partially system-dependent, but the sonic characteristics of a specific cable will work better musically in some systems than in others. Personal auditioning is the *only* way to evaluate cables and interconnects. Never be swayed by technical jargon about why one cable is better than another. Much of this is pure marketing hype, with little or no relevance to how the cable will perform musically in your system. Trust your ears.

That said, poorly designed cables—or cables not designed for audio, such as "lamp cord" sold in hardware stores—will degrade the sound of any system, and in similar ways.

Cables and interconnects can add some annoying distortions to the music. I've listed the most common sonic problems of cables and interconnects. (A full description of these terms is included in Chapter 4.)

Grainy and hashy treble Many cables overlay the treble with a coarse texture. The sound is rough rather than smooth and liquid.

Bright and metallic treble Cymbals sound like bursts of white noise rather than a brass-like shimmer. They also tend to splash across the soundstage rather than sounding like compact images. Sibilants (*s* and *sh* sounds on vocals) are emphasized, making the treble sound spitty. It's a bad sign if you suddenly notice more sibilance. The opposite condition is a dark and closed-in treble. The cable should sound open, airy, and extended in the treble without sounding overly bright, etched, or analytical.

Hard textures and lack of liquidity Listen for a glassy glare on solo piano in the upper registers. Similarly, massed voices can sound glazed and hard rather than liquid and richly textured.

Listening fatigue A poor cable will quickly cause listening fatigue. The symptoms of listening fatigue are headache, a feeling of relief when the music is turned down or stopped, the need to do something other than listen to music, and the feeling that your ears are tightening up. This last condition is absolutely the worst thing any audio component can do. Good cables (in a good system) will let you listen at higher levels for longer periods of time. If a cable or interconnect causes listening fatigue, avoid it no matter what its other attributes.

Lack of space and depth Using a recording with lots of natural depth and ambiance, listen for how the cable affects soundstage depth and the sense of instruments hanging in three-dimensional space. Poor cables can also make the soundstage less transparent.

Low resolution Some cables and interconnects sound smooth, but they obscure the music's fine detail. Listen for low-level information and an instrument's inner detail. The opposite of smoothness is a cable that's "ruthlessly revealing" of every detail in the music, but in an unnatural way. Musical detail should be audible, but not hyped or exaggerated. The cable or interconnect should strike a balance between resolution of information and a sense of ease and smoothness.

Mushy bass or poor pitch definition A poor-quality cable or interconnect can make the bass slow, mushy, and lacking in pitch definition. With such a cable, the bottom end is soggy and fat rather than taut and articulate. Low-frequency pitches are obscured, making the bass sound like a roar instead of being composed of individual notes.

Constricted dynamics Listen for the cable or interconnect's ability to portray the music's dynamic structure, on both small and large scales. For example, a guitar string's transient attack should be quick, with a dynamic edge. On a larger scale, orchestral climaxes should be powerful and have a sense of physical impact (if the rest of your system can portray this aspect of music).

Chapter 11

I must reiterate that putting a highly colored cable or interconnect in your system to correct a problem in another component (a dark-sounding cable on a bright loudspeaker) isn't the best solution. Instead, use the money you would have spent on new cables toward better loudspeakers—*then* go cable shopping. Cables and interconnects shouldn't be Band-Aids; instead, cables should be the finishing touch to let the rest of your components perform at their highest level.

Binding Posts and Cable Terminations

Binding posts vary hugely in quality, from the tiny spring-loaded, push-in terminal strips on cheap loudspeakers to massive, custom-made, machined brass posts plated with exotic metals. Poor binding posts not only degrade the sound, they also break easily. When shopping for power amplifiers and loudspeakers, take a close look at binding-post quality.

The most popular type is the five-way binding post. It accepts spade lugs, banana plugs, or bare wire. Some five-ways are nickel-plated; higher-quality ones are plated with gold and won't tarnish. Five-way binding posts should be tightened with a 1/2" nut driver, not a socket and ratchet or wrench that could easily overtighten the nut. The connection should be tight, but not to the point of stripping the post or causing it to turn in the chassis. When tightening a five-way binding post, watch the inside ring or collar next to the chassis; if it begins to turn, you've overtightened the post and are in danger of damaging the power amplifier or loudspeaker.

If you have a choice of bare wire, banana plug, or spade lug on loudspeaker cable terminations, go with the spade lug. It forms the best contact with a binding post and is the most standard form of connection.

Bi-Wired Loudspeaker Cables

Bi-wiring is running two lengths of cable between the power amplifier and loudspeaker. This technique usually produces better sound quality than conventional single-wiring. Most high-end loudspeakers have two pairs of binding posts for bi-wiring, with one pair connected to the crossover's tweeter circuit and the other pair connected to the woofer circuit. The jumpers connecting the two pairs of binding posts fitted at the factory must be removed for bi-wiring (Fig.11-1).

Fig.11-1 You must remove the jumper connecting the two sets of binding posts (as seen in the right pair of binding posts) for bi-wire operation.

In a bi-wired system, the power amplifier "sees" a higher impedance on the tweeter cable at low frequencies, and a lower impedance at high frequencies. The opposite is true in the woofer-half of the bi-wired pair. This causes the signal to be split up, with high frequencies traveling mostly in the pair driving the loudspeaker's tweeter circuit and low frequencies conducted by the pair connected to the loudspeaker's woofer circuit. This frequency splitting reportedly reduces magnetic interactions in the cable, resulting in better sound. The large magnetic fields set up around the conductors by low-frequency energy can't affect the transfer of treble energy. No one knows exactly how or why bi-wiring works, but on nearly all loudspeakers with bi-wiring provision, it makes a worthwhile improvement in the sound. Whatever your cable budget, you should bi-wire if your loudspeaker has bi-wired inputs, even if it means buying two runs of less expensive cables.

You can bi-wire your loudspeakers with two identical single-wire runs, or with a specially prepared bi-wire set. A bi-wire set has one pair (positive and negative) of terminations at the amplifier end of the cable, and two pairs at the loudspeaker end of the cable. This makes it easier to hook up, and probably offers slightly better sound quality.

Most bi-wired sets use identical cables for the high- and low-frequency legs. Mixing cables, however, can have several advantages. By using a cable with good bass on the low-frequency pair, and a more expensive but sweeter-sounding cable on the high-frequency pair, you can get better performance for a lower cost. Use a less expensive cable on the bass and put more money into the high-frequency cable. If you've already got two pairs of cable the same length, the higher-quality cable usually sounds better on the high-frequency side of the bi-wired pair. If you use different cables for bi-wiring, they should be made by the same manufacturer and have similar physical construction. If the cables in a bi-wired set have different capacitances or inductances, those capacitances and inductances change the loudspeaker's crossover characteristics.

Interconnects: Balanced and Unbalanced

Line-level interconnects come in two varieties: balanced and unbalanced. A balanced interconnect is recognizable by its three-pin XLR connector. An unbalanced interconnect is usually terminated with an RCA plug (Fig.11-2).

Fig.11-2 An RCA-terminated cable (left) carries an unbalanced audio signal. An XLR-terminated cable carries a balanced audio signal. (Courtesy Monster Cable Products)

Why do we use two incompatible systems for connecting equipment? At one time, all consumer audio hardware had unbalanced inputs and outputs, and all professional gear was balanced. In fact, balanced inputs are often called "professional inputs" to differentiate them from "consumer" unbalanced jacks. But what exactly is a balanced line, and how is it different from a standard RCA cable and jack?

In an unbalanced line, the audio signal appears across the center pin of the RCA jack and the shield, or ground wire. Some unbalanced interconnects have two signal conductors and a shield, with the shield not used as a signal conductor.

A balanced line has three conductors: two carrying signal, and one ground. The two signals in a balanced line are identical, but 180° out of phase with each other. When the signal in one of the conductors is at peak positive, the signal in the other conductor is at peak negative. The third conductor is signal ground. (Some balanced interconnects use three conductors plus a shield.)

The advantage of a balanced connection is that any noise picked up in the cable will be rejected by the component receiving the signal—a power amplifier, for example, when a balanced interconnect is run between those two components. This is why balanced lines are used almost exclusively in professional audio, particularly for very long runs that are subject to picking up noise.

In a high-end system, there's no clear-cut preference for using balanced or unbalanced connections. If you have a choice between connecting your equipment through balanced or unbalanced connections, try both and decide for yourself which sounds better (your dealer will often lend you interconnects). As in all things audio, the proof is in the listening. Let your ears decide if the component works best in your system when connected via the balanced or unbalanced lines.

Cable and Interconnect Construction

Cables and interconnects are composed of three main elements: the signal conductors, the dielectric, and the terminations. The *conductors* carry the audio signal; the *dielectric* is an insulating material between and around the conductors; and the *terminations* provide connection to audio equipment. These elements are formed into a physical structure called the cable's *geometry*. Each of these elements—particularly geometry—can affect the cable's sonic characteristics.

Conductors are usually made of high-purity copper wire. So-called "six-nines" copper is 99.9999% pure. Some cables and interconnects use silver-plated copper, or even pure silver conductors. The latter are extremely expensive, and have a characteristic sound.

The dielectric is the material surrounding the conductors, and is what gives cables and interconnects some of their bulk. The dielectric material has a large effect on the cable's sound; comparisons of identical conductors and geometry, but with different dielectric materials, demonstrate the dielectric's importance. Dielectric materials found in today's high-end cables include PVC, polyethylene, polypropylene, or even Teflon in the most expensive cables.

The terminations at the ends of cables and interconnects are part of the transmission path. High-quality terminations are essential to a good-sounding cable. We want a large surface contact between the cable's plug and the component's jack, and high contact pressure between them. RCA plugs will sometimes have a slit in the center pin to improve contact with the jack. Some RCA plugs have a locking mechanism that allows you to tighten the plug around the jack.

How all of these elements are arranged constitutes the cable's geometry. Some designers maintain that geometry is the most important factor in cable design—even more important than the conductor material and type. Geometry affects the magnetic interaction between the individual strands, among other factors.

Cable designers balance all these factors—conductor material, dielectric, and geometry—in an attempt to get the best-sounding cable possible.

Equipment Racks

You'll often see mid-fi audio gear housed in "stereo stands" with flimsy shelves. But to get the best performance from your high-end equipment, you'll need a solid, vibration-resistant platform for your components. A quality equipment rack does more than provide convenient housing for your equipment; it also isolates equipment from vibration. There is no question that vibrations degrade the sonic performances of preamplifiers, CD players, and particularly turntables. This vibration is generated by transformers in your components' power supplies, motors in turntables and CD players, and from acoustic energy impinging on the electronics. Yes, racks can have an audible effect on a system.

A good equipment rack fights vibration with rigidity, mass, and careful design. The massive, inert structure of a high-quality rack is much less likely to vibrate when in the presence of sound pressure generated by the loudspeakers. Moreover, the equipment rack can absorb, or damp, the vibration created by power transformers and motors. Many equipment racks have built-in vibration-damping mechanisms in their shelves.

Many equipment racks are supplied with spikes to couple the stand to the floor and "drain" vibration from the rack. For a spiked rack to be effective, the floor must be sturdy and flat so that the rack doesn't rock. The heights of most spikes are adjustable; you can level the rack and get good contact between the floor and all four (or three) spikes. Rack spikes are usually much heavier duty, with more rounded points than loudspeaker spikes. Note, however, that both types of spike will damage wood floors. Some racks are supplied with small plates to hold the spikes, preventing floor damage.

A rack's performance can often be improved by adding aftermarket spikes, cones, or other vibration-isolation devices underneath your components. These devices further decouple the audio component from the rack and sometimes deliver a significant sonic improvement for a relatively small investment.

Avoid racks with large, unsupported shelves, flimsy construction, low mass, and poor vibrational damping. And don't even consider the generic "stereo stands" sold in furniture and department stores. Get a rack specially designed for high-performance audio systems. Consider a good equipment rack an essential part of your hi-fi system; it will help your system achieve its full musical potential.

You'll need to decide whether to house your equipment in an open-air rack or stand (Fig.11-3), or behind closed doors in a cabinet. The open-air rack is the preferred choice of audiophiles—it allows easier access to the connections, affords better ventilation of heat-generating components, and showcases your equipment.

Fig.11-3 A high-quality equipment rack provides a stable, vibration-resistant platform for your components and keeps them well ventilated. (Courtesy Music Direct)

AC Power Conditioners

Once your equipment is housed in a rack or enclosed cabinet, you'll need to plug the gear into an AC power source. Rather than plug the components into an AC wall socket, you should invest in an AC power conditioner. An AC power conditioner plugs into the wall outlet and provides multiple AC outlets for plugging in your audio equipment (Fig.11-4). An AC power conditioner is the single most important "accessory" you can add to your system. In fact, I consider an AC conditioner not just an accessory, but an essential component of any high-performance audio system.

Fig.11-4 An AC power conditioner provides enough outlets for your system, protects your equipment from lightning, and improves the sound quality. (Courtesy Monster Cable Products)

AC conditioners act in two ways to improve the sound of an audio system. First, AC conditioners filter noise from the AC line before it gets to your audio components. This noise on the AC power line is generated by light dimmers, refrigerators, motors, and household appliances. Industrial motors connected to the power grid also pollute the AC line with hash and high-frequency noise. This noise gets into the audio signal and degrades the sound.

A second source of dirty AC is your equipment itself. Any component using a microprocessor or other digital circuits (all digital source components, A/V controllers and receivers, and even some analog preamplifiers) put noise on the power line through their AC power-line cords. This noise then gets into your other components and reduces sound quality. The AC ground connects all the chassis of an audio system. If you've got a noisy ground on one component, you've got a noisy ground on all your components. For example, digital noise in a CD player's ground can get into your pre-amplifier, with the AC power line acting as a conduit for this noise.

All of these problems can be controlled with a well-designed AC power-line conditioner. First, nearly all conditioners filter the incoming AC line to remove the high-frequency garbage generated by factories, neighbors, and your own appliances. The filters allow the 60Hz AC to pass, but remove noise from the line. Second, some filters isolate the components from each other with small isolation transformers on some of the conditioner's AC outlets. These transformers break the physical connection between components, preventing noise from traveling from one component to another. The isolated outputs are often marked "digital" for plugging in digital components, preventing a CD player from polluting the AC supplying the preamplifier, for example. Third, a good line conditioner will reduce the amount of noise coupled to signal ground. Finally, AC line conditioners can protect components from voltage spikes, lightning strikes, and surges in the power-supply voltage. Not all conditioners perform every function listed here; conditioners vary in their design principles, with some addressing one problem but not another.

When choosing a line conditioner, make sure its power capability exceeds the power consumption of the components you'll be plugging into it. Each component's owner's manual will state the component's power consumption in watts. Add together the individual power-consumption specs to determine the total amount of power to be drawn from the conditioner. Compare this number with the power conditioner's maximum rated power delivery. Also look for the UL (Underwriters Laboratories) or CSA (Canadian Standards Association) seal of approval, indicating that the power conditioner meets certain safety requirements. Choose a conditioner with a sufficient number of outlets for your present and anticipated needs. As with all accessories, try the power conditioner in your system before you buy. Expect to pay a minimum of $200 for a conditioner with just a few outlets, to several thousand dollars for a state-of-the-art system. Many excellent conditioners cost less than $300.

A power-line conditioner can't make poor audio components sound good; instead, it merely provides the optimum AC environment for those components so that they may realize their full potentials. The sonic benefits of a good line conditioner include a "blacker" background, with less low-level grunge and noise. The music seems to emerge from a perfectly quiet and black space, rather than a grayish background. The treble often becomes sweeter, less grainy, and more extended. Soundstaging often improves, with greater transparency, tighter image focus, and a newfound soundstage depth. Midrange textures become more liquid, and the presentation has an ease and musicality not heard without the conditioner.

12

How to Get the Best Sound From Your System

How an audio system is set up has an enormous effect on the system's sound quality. In fact, system set-up is an art that can turn a mediocre-sounding system into one that sounds spectacular. A budget system that is expertly set-up will often outperform a more costly one that has not been assembled with such expertise.

I'm fortunate to have learned the art of system set-up from the world's great experts—the designers of high-end audio products. In the course of my work as a full-time equipment reviewer since 1989, I've had a succession of designers bring their products to my listening room for review. As you can imagine, the designer or company representative is highly motivated to make his product perform at its best for the review. From each of them I've learned different techniques for squeezing the most performance out of an audio system. I'll share with you these "tricks of the trade."

In this chapter, I'll guide you through the steps in setting up a system and tuning it for the best possible sound. The techniques presented in this chapter should be used in conjunction with the critical listening criteria described in Chapter 3. You will often make a change to the system, listen to evaluate the results, and make another change in a repetitive process until the system has achieved its full potential.

We'll also spend quite a bit of time in this chapter on room acoustics and how you can use common domestic materials such as rugs and bookcases to improve your room's sound.

Although it's enormously rewarding to turn a decent-sounding system into a great-sounding one using only your skill and knowledge, such tweaking isn't for everyone. Many music lovers would rather just enjoy the music and avoid the hands-on aspects of system set-up. For those readers, there's no substitute for the help of a skilled and caring dealer. Most dealers will come to your home and install a system you've purchased from them, and the good ones will spend the time to coax the best performance from it.

Room Layout

Your first decision in system set-up is where to put the equipment, the loudspeakers, and the listening chair or couch. If your audio system is also used for home theater, many of these factors will have been decided by your television's location. Although this chapter focuses on left- and right-channel speaker placement and room acoustics, that basic set-up should serve as a foundation for adding a center-channel speaker, a subwoofer, and surround speakers for film-soundtrack reproduction or multichannel music. More detail on surround-sound set-up is included in Chapter 10.

Chapter 12

Loudspeaker Placement

To hear the full magic of a high-quality audio system, you'll need to arrange your listening room in a way that allows the system to perform at its best. All the effort you've put into choosing a system (along with the money you spent on it) can be wasted without a correct fundamental setup. Positioning the speakers in roughly a triangle with the listener (more on this later) will get you in the ballpark and give your system a chance of being fine-tuned for optimum performance. At the next level, small loudspeaker movements within this fundamentally correct placement zone allow you to precisely dial-in the system.

Finding the right spot for your loudspeakers is the single most important factor in getting good sound. Loudspeaker placement affects tonal balance, the quantity and quality of bass, soundstage width and depth, midrange clarity, articulation, and imaging. As you make large changes in loudspeaker placement, then fine-tune placement with smaller and smaller adjustments, you'll hear a newfound musical rightness and seamless harmonic integration to the sound. When you get it right, your system will come alive. Best of all, it costs no more than a few hours of your time.

Before getting to specific recommendations, let's cover the six fundamental factors that affect how a loudspeaker's sound will change with placement. (Later we'll look at each of these factors in detail.) Note that you should wait until after you've completed the entire loudspeaker placement procedure to install the loudspeaker's floor-coupling spikes.

1) The relationship between the loudspeakers and the listener is of paramount importance. The listener and loudspeakers should form a triangle; without this basic setup, you'll never hear good soundstaging and imaging.

2) Proximity of loudspeakers to walls affects the amount of bass. The nearer the loudspeakers are to walls and corners, the louder the bass.

3) The loudspeaker and listener positions in the room affect the audibility of room resonant modes. Room resonant modes are reinforcements and cancellations at certain frequencies that create peaks and dips in the frequency response, which can add an unnatural "boominess" to the sound. When room resonant modes are less audible, the bass is better defined, and midrange clarity increases.

4) The farther out into the room the loudspeakers are, the better the soundstaging—particularly depth.

5) Listening height affects tonal balance.

6) Toe-in (angling the loudspeakers toward the listener) affects tonal balance (particularly the amount of treble), soundstage width, and image focus.

Let's look at each of these factors in detail.

How to Get the Best Sound From Your System

1) Relationship between the loudspeakers and the listener

The most important factor in getting good sound is the geometric relationship between the two loudspeakers and the listener (we aren't concerned about the room yet). The listener should sit exactly between the two loudspeakers, at a distance away from each loudspeaker slightly greater than the distance between the loudspeakers themselves. Though this last point is not a hard-and-fast rule, you should certainly sit exactly between the loudspeakers; that is, the same distance from each one. If you don't have this fundamental relationship, you'll never hear good soundstaging from your system.

Fig.12-1 shows how your loudspeaker and listening positions should be arranged. The listening position—equidistant from the speakers, and slightly farther from each speaker than the speakers are from each other—is called the "sweet spot." This is roughly the listening position where the music will snap into focus and sound the best. If you sit to the side of the sweet spot, the soundstage will tend to bunch up around one speaker. This bunching-up effect will vary with the loudspeaker; some loudspeakers produce a wider sweet spot than others. When you sit exactly between the speakers, a "phantom" center image is created; you hear a vocalist, for example, as coming from a position between the speakers. When you move off to the side, the vocalist's image moves toward the speaker you are closer to.

Sweet Spot

Fig.12-1 The "sweet spot" is the listening position where the soundstage snaps into focus.

Setting the distance between the loudspeakers is a trade-off between a wide soundstage and a strong center image. The farther apart the loudspeakers are (assuming the same listening position), the wider the soundstage will be. As the loudspeakers are moved farther apart, however, the center image weakens, and can even disappear. If the loudspeakers are too close together, soundstage width is constricted.

The ideal speaker separation will produce a strong center image and a wide soundstage. There will likely be a position where the center image snaps into focus, appearing as a stable, pinpoint spot exactly between the loudspeakers. A musical selection with a singer and sparse accompaniment is ideal for setting loudspeaker spacing and ensuring a strong center image. With the loudspeakers fairly close together, listen for a tightly focused image exactly between the two loudspeakers. Move the loudspeakers a little farther apart and listen again. Repeat this move/listen procedure until you start to hear the central image become larger, more diffuse, and less focused, indicating that you've gone slightly beyond the maximum distance your loudspeakers should be from each other for a given listening position.

Note that adding a center-channel speaker for multichannel music or home theater reproduction makes the left-right placement less critical. The center speaker "fills in" the soundstage center, as well as broadens the sweet spot. This so-called "hard" center channel is contrasted with the "phantom" center channel created by your brain. When sitting between two speakers, you hear images coming from between the speakers, just as though a speaker were there. The phantom center channel is more fragile (the speaker and listener locations must be just right), but just as convincing. In addition, the vast majority of music recordings are in two-channel stereo; creating a multichannel mix from a two-channel recording involves signal processing (Dolby Pro Logic IIx or DTS Neo:6 Music, for examples) that many audiophiles are unwilling to use. This signal processing is provided only in home-theater controllers and A/V receivers, not in two-channel preamplifiers. Also consider that those two-channel recordings were created over a two-channel system and were meant to be played back over two loudspeakers.

A factor to consider in setting this angle is the relationship to the room. You can have the same geometric relationship between loudspeakers and listener with the loudspeakers close together and a close listening position, or with the loudspeakers far apart and a distant listening position. At the distant listening position, the listening room's acoustic character will affect the sound more than at the close listening position. That's because you hear more direct sound from the loudspeaker and less reflected sound from the room's walls. Consequently, the farther away you sit, the more spacious the sound. The closer you sit, the more direct and immediate the presentation. Some loudspeakers need a significant distance between the loudspeaker and the listener to allow the loudspeakers' individual drive units to integrate. If you hear a large tonal difference just by sitting closer, you should listen from a point farther away from the speakers.

2) Proximity to walls affects the amount of bass

The room boundaries have a great effect on a loudspeaker's overall tonal balance. Loudspeakers placed close to walls will exhibit a reinforcement in the bass (called "room gain"), making the musical presentation weightier. Some loudspeakers are designed to be near a rear wall (the wall behind the speakers); they need this reinforcement for a natural tonal balance. These loudspeakers sound thin if placed out into the room. Others sound thick and heavy if not at least 3' from the rear and side walls. Be aware of which type you're buying if your placement options are limited.

When a loudspeaker is placed near a wall, its bass energy is reflected back into the room essentially in phase with the loudspeaker's output. This means the direct and reflected waves reinforce each other at low frequencies, producing louder bass. The closer to the corners the loudspeakers are placed, the more bass you'll hear.

The loudspeaker's position in relation to the rear and side walls will also affect which frequencies are boosted. Correct placement can not only extend a loudspeaker's bass response by complementing its natural rolloff, but also avoid peaks and dips in the response. Improper placement can cause frequency-response irregularities that color the bass. That is, some frequencies are boosted relative to others, making the bass reproduction less accurate. For this reason, the loudspeakers should be positioned at different distances from the rear and side walls. A rule of thumb: the two distances should not be within 33% of each other. For example, if the loudspeaker is 3' from the side wall, it should also be at least 4' from the rear wall.

3) Loudspeaker and listener positions affect room-mode audibility

In addition to deepening bass extension and smoothing bass response, correct loudspeaker placement in relation to the room's walls can also reduce the audible effects of your room's resonant modes. Room resonant modes are reinforcements and cancellations at certain frequencies that create peaks and dips in the frequency response. In addition, loudspeaker and listener placement affect *standing waves*, which are stationary patterns of high and low sound pressure in the room that color the sound. The standing-wave patterns in a room are determined both by the room's dimensions and by the position of the sound source in the room. By putting the loudspeakers and listener in the best locations, we can achieve smoother bass response.

A well-known rule of thumb states that, for the best bass response, the distance between the loudspeakers and the rear wall should be one-third of the length of the room (Fig.12-2). If this is impractical, try one-fifth of the room length. Both of these positions reduce the excitation of standing waves and help the loudspeaker integrate with the room. Starting with these basic configurations, move the loudspeakers and the listening chair in small increments while playing music rich in low frequencies. Listen for smoothness, extension, and how well the bass integrates with the rest of the spectrum. When you find the loudspeaker placement where the bass is the smoothest, you should also hear an increase in midrange clarity and definition.

You should also experiment with moving the listening seat forward and backward to adjust the amount of bass you hear. If the sound is boomy, you could be sitting in a standing wave peak; if the sound is thin, the listening seat could be in a standing wave null. The solution is to move the listening seat to achieve just the right bass balance.

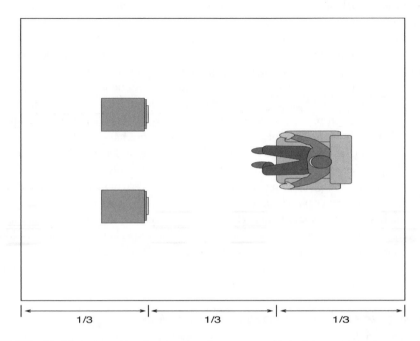

Fig.12-2 The ideal listener and loudspeaker locations are one-third of the way into the room. If this is impractical, one-fifth of the way into the room is the next best choice.

4) Distance from rear wall affects soundstaging

Generally, the farther away from the rear wall the loudspeakers are, the deeper the soundstage. A deep, expansive soundstage is rarely developed with the loudspeakers near the rear wall. Pulling the loudspeakers out a few feet can make the difference between poor and spectacular soundstaging. Unfortunately, many living rooms don't accommodate loudspeakers far out into the room.

5) Listening height and tonal balance

Most loudspeakers exhibit changes in frequency response with changes in listening height. These changes affect the midrange and treble, not the bass balance. Typically, the loudspeaker will be brightest (i.e., have the most treble) when your ears are at the same height as the tweeters, or on the tweeter axis. Most tweeters are positioned between 32" and 40" from the floor to coincide with typical listening heights. If you've got an adjustable office chair, you can easily hear the effects of listening axis on tonal balance.

The degree to which the sound changes with height varies greatly with the loudspeaker. Some models have a broad range over which little change is audible; others can exhibit large tonal changes when you merely straighten your back while listen-

ing. Choosing a listening chair that sets your ears at the optimum axis will h
a good treble balance.

6) Toe-in

Toe-in is pointing the loudspeakers inward toward the listener rather than facing them straight ahead (see Fig.12-3). There are no rules for toe-in; the optimum amount varies greatly with the loudspeaker and the listening room. Some loudspeakers need toe-in; others work best firing straight ahead. Toe-in affects many aspects of the musical presentation, including mid- and high-frequency balance, soundstage focus, sense of spaciousness, and immediacy.

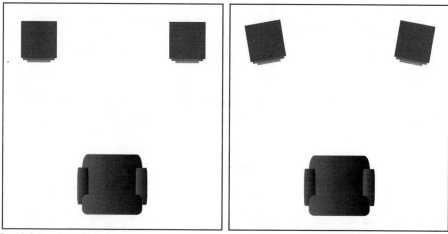

Fig.12-3 Loudspeakers can be positioned with no toe-in (left) or with toe-in (right).

Most loudspeakers sound the brightest directly on-axis (directly in front of the loudspeaker). Toe-in therefore increases the amount of treble heard at the listening seat. An overly bright loudspeaker can often be tamed by pointing the loudspeaker straight ahead. Some models, designed for listening without toe-in, are far too bright on-axis.

A toed-in loudspeaker will present more direct energy to the listener and project less energy into the room, where it might reach the listener only after reflecting from room surfaces. As we'll see later in this chapter, sound reflected from the sidewalls to the listening positions can degrade sound quality. Toe-in often increases soundstage focus and image specificity. When toed-in, many loudspeakers provide a more focused and sharply delineated soundstage. Images are more clearly defined, compact, and tight, rather than diffuse and lacking a specific spatial position. The optimum toe-in is often a trade-off between too much treble and a strong central image. With lots of toe-in, the soundstage snaps into focus, but the presentation is often too bright. With no toe-in, the treble balance is smoother, but the imaging is more vague.

Toe-in also affects the presentation's overall spaciousness. No toe-in produces a larger, more billowy, less precise soundstage. Instruments are less clearly delineated, but the presentation is bigger and more expansive. Toeing-in the loudspeakers shrinks the apparent size of the soundstage, but allows more precise image localization. Again, the proper amount of toe-in depends on the loudspeaker, room, and personal preference. There's no substitute for listening, adjusting toe-in, and listening again.

Identical toe-in for each loudspeaker is vital. This is most easily accomplished by measuring the distances from the rear wall to each of the loudspeaker's rear edges; these distances will differ according to the degree of toe-in. Repeat this procedure on the other loudspeaker, adjusting its toe-in so that the distances match those of the first loudspeaker. Another way to ensure identical toe-in is to sit in the listening seat and look at the loudspeakers' inside edges. You should see the same amount of each loudspeaker cabinet's inner side panel. Identical toe-in is essential to soundstaging because the speaker's frequency response at the listening position changes with toe-in, and hearing the identical frequency response from each speaker is an important contributor to precise image placement within the soundstage.

Keep in mind that all loudspeaker placement variations are interactive with one another, particularly toe-in and the distance between loudspeakers. For example, a wide soundstage can be achieved with narrow placement but no toe-in, or wide placement with extreme toe-in.

Dipolar and Bipolar Loudspeaker Placement

Dipolar loudspeakers produce sound to the rear as well as to the front. An electrostatic speaker is a dipole because the vibrating diaphragm sits in open space rather than in a cabinet, launching sound equally to the front and rear. This rear wave from dipolar speakers is out of phase with the front wave; that is, when the diaphragm moves forward to create positive pressure in front of the diaphragm, it creates negative pressure behind the diaphragm.

A bipolar speaker typically uses arrays of conventional dynamic drivers on the front and rear of the loudspeaker enclosure. The front and rear waves from a bipolar speaker are *in phase* with each other. That's the difference between dipolar and bipolar: a dipole's rear wave is *out of phase* with the front wave, and a bipole's rear wave is *in phase* with the front wave.

Both of these speaker types are covered in detail in Chapters 9 and 10. What concerns us here, however, are the special placement requirements of bipolar and dipolar loudspeakers, along with the different ways in which they interact with the listening room.

The most important consideration when positioning dipoles is that the wall behind the speakers (the rear wall) has a much greater influence on the sound than it does with conventional point-source speakers (those that direct energy in only one direction). Conversely, how the sidewalls are treated is less important with dipoles because they radiate very little energy to the sides. (See Fig.9-5, which shows the dispersion patterns of point-source and dipolar speakers.)

Generally, dipoles like a reflective rear wall, but with some diffusing objects behind the speaker to break up the reflected energy. A highly absorbent rear wall defeats the purpose of a dipole; that reflected energy is beneficial, and you want to hear it. But if the wall is flat and lacks surfaces that scatter sound, the reflected energy combines with the direct sound in a way that reduces soundstage depth. Bookcases directly behind dipolar speakers help diffuse (scatter) the rear wave, as do rock fireplaces, furniture, and other objects of irregular shape.

Dipolar loudspeakers also need to be placed farther out into the room than conventional point-source speakers. You can't put dipoles near the rear wall and expect a big, deep soundstage. Be prepared to give up a significant area of your listening room to dipolar speakers.

Subwoofer Placement and Setup

It's relatively easy to put a subwoofer into your system and hear more bass. What's difficult is making the subwoofer's bass integrate with the sound of your main speakers. Low bass as reproduced by a subwoofer's big cone can sound different from the bass reproduced by the smaller cones in the left and right speakers. A well-integrated subwoofer produces a seamless sound, no boomy thump, and natural reproduction of music. A poorly integrated subwoofer will sound thick, heavy, boomy, and unnatural, calling attention to the fact that you have smaller speakers reproducing the frequency spectrum from the midrange up, and a big subwoofer putting out low bass.

Integrating a subwoofer into your system is challenging because the main speakers may have small cones, and the subwoofer has a large and heavy cone. Moreover, the subwoofer is optimized for putting out lots of low bass, not for reproducing detail. The main speakers' upper bass is quick, clean, and articulate. The subwoofer's bass is often slow and heavy.

Achieving good integration between small speakers and a subwoofer is easier if you buy a complete system made by one manufacturer. Such systems are engineered to work together to provide a smooth transition between the subwoofer and the main speakers. Specifically, the crossover network removes bass from the left and right speakers, and removes midrange and treble frequencies from the signal driving the subwoofer. If all these details are handled by the same designer, you're much more likely to get a smooth transition than if the subwoofer is an add-on component from a different manufacturer.

If you do choose a subwoofer made by a different manufacturer, several controls found on most subwoofers help you integrate the sub into your system. One control lets you adjust the *crossover frequency*. This sets the frequency at which the transition between the subwoofer and the main speakers takes place. Frequencies below the crossover frequency are reproduced by the subwoofer; frequencies above the crossover frequency are reproduced by the main speakers. If you have small speakers that don't go very low in the bass and you set the crossover frequency too low, you'll get a "hole" in the frequency response. That is, there will be a narrow band of frequencies that aren't reproduced by the woofer *or* the main speakers. In a two-channel music system, the crossover will likely be inside the woofer. In a system designed for home theater as well as music, the crossover will be inside the A/V controller or A/V receiver.

Setting the subwoofer's crossover frequency too high also results in poor integration, but for a different reason. The big cone of a subwoofer is specially designed to reproduce low bass. When it is asked to also reproduce upper-bass frequencies, those upper-bass frequencies are less clear and distinct than if they were reproduced by the smaller main speakers. Finding just the right crossover frequency is the first step in achieving good integration. Most subwoofer owner's manuals include instructions for setting the crossover frequency. As a rule of thumb, the lower the subwoofer's crossover is set, the better.

Some subwoofers also provide a knob or switch marked *Phase*. To understand a subwoofer's phase control, visualize a sound wave being launched from your subwoofer and from your main speakers at the same time. Unless the main speakers and subwoofer are identical distances from your ears, those two sound waves will reach your ears at different times, or have a phase difference between them. In addition, the electronics inside a subwoofer can create a phase difference in the signal. The subwoofer's phase control lets you delay the wave generated by the subwoofer so that it lines up in time with the wave from the main speaker. When the sound waves are in-phase, you hear a more coherent and better-integrated sound.

One way of setting the phase control is to sit in the listening position with music playing through the system. Have a friend rotate the phase control (or flip the phase switch) until the bass sounds the smoothest.

But there's a much more precise way of setting the phase control that guarantees perfect phase alignment between the subwoofer and main speakers. First, reverse the connections on your main loudspeakers so that the black speaker wire goes to the speaker's red terminal, and the red speaker wire goes to the speaker's black terminal. Do this with both speakers. Now, from a test CD that includes pure test tones, select the track whose frequency is the same as the subwoofer's crossover frequency. Sit in the listening position and have a friend rotate the subwoofer's phase control until you hear the *least* amount of bass. The subwoofer's phase control is now set perfectly. Return your speaker connections to their previous (correct) positions: red to red, black to black.

Here's what's happening when you follow this procedure: By reversing the polarity of the main speakers, you're putting them out of phase with the subwoofer. When you play a test signal whose frequency is the same as the subwoofer's crossover point, both the sub and the main speakers will be reproducing that frequency. You'll hear minimum bass when the waves from the main speakers and subwoofers are maximally out of phase. That is, when the main speaker's cone is moving in, the subwoofer's cone is moving out. The two out-of-phase waves cancel each other, producing very little bass. Now, when you return your loudspeakers to their proper connection (putting them back in-phase with the subwoofer), they will be maximally *in-phase* with the subwoofer. This is the most accurate method of setting a subwoofer's phase control because it's much easier to hear the null rather than the peak. Unless you move the subwoofer or main speakers, you need to perform this exercise only once.

You can also get more dynamic impact and clarity from your subwoofer by placing it close to the listening position. Sitting near the subwoofer causes you to hear more of the sub's direct sound and less of the sound after it has been reflected around the room. You hear—and feel—more of the low-frequency wave launch, which adds to the visceral impact of owning a subwoofer. Bass impact is more startling, powerful, and dynamic when the subwoofer is placed near the listening position.

Subwoofer placement also has a large effect on how much bass you hear and how well the sub integrates with your main speakers. When a subwoofer is correctly positioned, the bass will be clean, tight, quick, and punchy. A well-located subwoofer will also produce a seamless sound between the sub and the front speakers; you won't hear the subwoofer as a separate speaker. A poorly positioned subwoofer will sound boomy, excessively heavy, thick, lacking detail, slow, and have little dynamic impact. In addition, you'll hear exactly where the front speakers leave off and the subwoofer takes over.

Some general guidelines for subwoofer placement: As with full-range speakers, avoid putting the subwoofer the same distance from two walls. For example, if you have a 20'-wide room, don't put the subwoofer 10' from each wall. Similarly, don't put the subwoofer near a corner and equidistant from the side and rear walls. Instead, stagger the distances to each wall. Staggering the subwoofer's distance from each wall smoothes the bass because the frequencies being reinforced by the wall are randomized rather than coincident.

Multichannel Loudspeaker Placement

So far, we've discussed the placement of two loudspeakers for stereo music reproduction. With multichannel music and home theater becoming increasingly common, let's expand on these loudspeaker-placement principles to include positioning more than two loudspeakers.

The optimum loudspeaker radiation pattern (uni-polar, dipolar) and configuration differ for multichannel music reproduction and home theater. For multichannel music, the ideal loudspeaker array is five identical full-range loudspeakers placed equidistant from the listener. For film-soundtrack reproduction, the center loudspeaker is typically smaller and lacks bass extension, and the surround speakers are dipolar types mounted on the sidewalls. Loudspeaker arrays optimized for home theater also include a subwoofer.

Let's first look at the ideal multichannel music loudspeaker array. The center loudspeaker should be positioned on the room's center-line directly in front of the listening position, and slightly behind the plane of the left and right loudspeakers. This placement creates a gentle arc, and puts the center loudspeaker at the same distance from the listener as the left and right loudspeakers. If the three front loudspeakers were lined up, the sound from the center speaker would reach the listener before the sound from the left and right speakers.

Chapter 12

The rear-channel speakers should be located at 135–150° as shown in Fig.12-4 and at the same distance from the listener as the front three loudspeakers. This placement isn't always practical, however, so many multichannel products provide a rear-channel delay for those situations in which the listener must sit closer to the rear loudspeakers. Delaying the signals to the rear channels causes the sound from the rear speakers to reach the listener at the same time as sound from the front speakers.

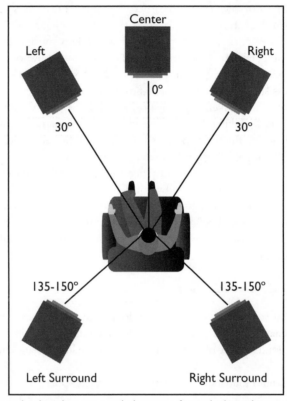

Fig.12-4 The optimum loudspeaker array and placement for multichannel music reproduction.

With this array, the front left and rear left loudspeakers can produce phantom images between them along the left sidewall, and the front right and rear right loudspeakers create phantom images along the right sidewall. Correct loudspeaker placement helps to achieve a soundfield that appears to be continuous from front to back, rather than as two separate soundfields at the front and rear of the room.

Loudspeaker arrays optimized for film-soundtrack reproduction usually employ dipolar surround loudspeakers rather than point-source loudspeakers. Dipolar speakers produce sound equally to the front and rear; when positioned on the sidewalls, the listener hears sound from the surround speakers only after it has been reflected from the room's boundaries. This simulates the array of multiple surround speakers in a movie theater from just two surround speakers, and creates a greater feeling of envelopment. (Chapter 10 includes more detailed information on multichannel loudspeaker placement for home theater.)

Loudspeaker Placement Summary

Loudspeaker placement is the single most important thing you can do to improve your system's sound. It's free, helps develop listening skills, and can make the difference between mediocre and spectacular sound with the same electronics and loudspeakers. Before spending money on upgrading components or acoustic treatments, be sure you've realized your system's potential with correct loudspeaker placement.

After you've found the best loudspeaker placement, install the carpet-piercing spikes (if any) supplied by the manufacturer. Level the spikes so that the loudspeaker doesn't rock: the loudspeaker's weight should be carried by all four (or three) spikes. If you have wood floors that you don't want to mar with spikes, place the round metal discs that are often supplied with the loudspeakers beneath the spikes.

You've seen how loudspeaker placement gives you precise and independent control over different aspects of the music presentation. You can control both the quantity and the quality of the bass by changing the loudspeakers' distances from the rear and side walls. The audibility of room resonance modes can be reduced by finding the best spots for the loudspeakers and listening chair. Treble balance can be adjusted by listening height and toe-in. The balance between soundstage focus and spaciousness is easily changed just by toeing-in the speakers. Soundstage depth can be increased by moving the speakers farther out into the room.

I've had the privilege of watching some of the world's greatest loudspeaker designers set up loudspeakers in my listening room for review. At the very highest levels of the art, tiny movements—half an inch, for example—can make the difference between very good and superlative sound. The process can take as little as two hours, or as long as three days. I've often had the experience of thinking the sound was excellent after half a day of moving the loudspeakers, only to discover that the loudspeaker was capable of much greater performance when perfectly dialed-in.

Loudspeaker positioning is a powerful tool for achieving the best sound in your listening room, and it doesn't cost a cent. Take advantage of it.

Optimizing Your Listening Room

The room in which music is reproduced has a profound effect on sound quality. In fact, the listening room's acoustic character should be considered another component in the playback chain. Because every listening room imposes its own sonic signature on the reproduced sound, your system can sound its best only when given a good acoustical environment. An excellent room can help get the most out of a modest system, but a poor room can make even a great system sound mediocre.

Chapter 12

Common Room Problems and How to Treat Them

Treating your listening room can range from simply hanging a rug on a wall to installing specially designed acoustic devices. Large gains in sound quality can be realized just by adding—or moving—common domestic materials such as carpets, area rugs, and drapes. This approach is inexpensive, simple, and often more aesthetically pleasing than installing less familiar acoustic products.

Here are some of the most common room problems, and how to correct them.

1) Untreated parallel surfaces

Perhaps the most common and detrimental room problems is that of untreated parallel surfaces. If two reflective surfaces face each other, flutter echo will occur. Flutter echo is a "pinging" sound that remains after the direct sound has stopped. If you've ever been in an empty, uncarpeted house and clapped your hands, you've heard flutter echo. It sounds like a ringing that hangs in the air long after the clap has decayed.

Flutter echo is easy to prevent. Simply identify the reflective parallel surfaces and put an absorbing or diffusing material on one of them. This will break up the repeated reflections between the surfaces. The material could be a rug hung on a wall, a carpet on the floor (if the flutter echo is between a hard floor and ceiling), drapes over a window, or an acoustically absorbent material applied to a wall.

2) Uncontrolled floor and side-wall reflections

It is inevitable that loudspeakers will be placed next to the room's side walls and near the floor. Sound from the loudspeakers reaches the listener directly, in addition to being reflected from the side walls, floor, and ceiling. Side-wall reflections are the music signal delayed in time, colored in timbre, and spatially positioned at different locations from the direct sound. All these factors can degrade sound quality. Moreover, floor and side-wall reflections interact with the direct sound to further color the music's tonal character. Fig.12-5 shows how the sound at the listening seat is a combination of direct and reflected sound.

Side-wall reflections color the music's tonal balance in three ways. First, virtually all loudspeakers' *off-axis responses* (frequency response measured at the side of the loudspeaker) are much less flat (accurate) than their on-axis responses. The sound emanating from the loudspeaker sides (the signal that reflects off the side wall) may have large peaks and dips in its frequency response. When this colored signal is reflected from the side wall to the listener, we hear this tonal coloration imposed on the music. Second, the side wall's acoustic characteristics will further color the reflection. If the wall absorbs high frequencies but not midband energy, the reflection will have even less treble.

Finally, when the direct and reflected sounds combine, the listener hears a combination of the direct sound from the loudspeaker and a slightly delayed version of

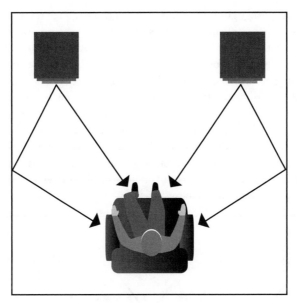

Fig.12-5 The sound at the listening position is a combination of direct and reflected sound.

the sound reflected from the side wall. The result is a phenomenon called _comb filtering_, a sequence of peaks and notches (hence its similarity to a comb) in the frequency response caused by constructive and destructive interference between the direct and reflected sounds. It all adds up to coloration of the signal at the listening position.

The result of these mechanisms—the loudspeaker's colored off-axis response, the sidewall's acoustic properties, and comb filtering—is a sound with a very different tonal balance from that of the direct signal from the loudspeakers. Side-wall reflections are one reason the same loudspeakers sound different in different rooms.

Fortunately, treating side-wall reflections is simple: just put an absorbing or diffusing material on the side walls between the loudspeakers and the listening position. Drapes are highly effective, particularly those with heavy materials and deep folds. The floor reflection is even easier to deal with: carpet or a heavy area rug on the floor will absorb most of the reflection and reduce its detrimental effects.

Diffusion turns the single discrete reflection into many lower-amplitude reflections spread out over time and reflected in different directions (see Fig.12-6). Diffusion can be achieved with specialized acoustic diffusers or with any irregular surface. A open-backed bookcase full of books makes an excellent diffuser, particularly if the books are of different depths, or are arranged with their spines sticking out at different distances.

Note that it isn't necessary to treat a listening room's entire side-wall area; the reflections come only from small points along the wall. At mid- and high frequencies, sound waves behave more like rays of light. We can thus trace side-wall reflections to the listening seat and put the absorber (drapes or hanging rug) or diffuser (bookcase or

CD racks) in exactly the right location. As with light rays, a sound wave's angle of inci-
dence equals its angle of reflection. That is, the angle at which a sound wave strikes a
reflective surface is the same as the angle at which it bounces off that surface. We can
exploit this fact to find the most effective position for the absorber or diffuser. Simply
sit in the listening seat and have a friend hold a hand mirror against the sidewall at the
level of the speaker's tweeter. Have him slowly move the mirror toward the back of
the room until you see the speaker's tweeter in the mirror. This point of visual reflection is also the point of acoustic reflection. Repeat the process for the other side wall.
If your listening room is symmetrical and the listening position is in the middle of the
room, you need use this technique on only one side wall, and then duplicate the
acoustic treatment on the other. To maintain acoustical symmetry in the room, both
side-wall treatments should be the same.

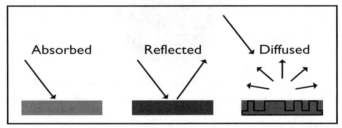

Fig.12-6 Sound striking a surface is absorbed, reflected, or diffused (or a combination of all three).

3) Thick, boomy bass

Thick, boomy bass is a common affliction that can be difficult to control. It often
results from room resonance modes, poor loudspeaker or listening-chair placement,
poor loudspeakers, or not enough low-frequency absorption in the listening room. As
we will see in the later section on standing waves, listening-seat position can also
increase bass bloat.

     If thick and boomy bass persists even after minimizing it with careful loudspeaker
placement (the most effective method of alleviating the problem), you may want to consider different loudspeakers. If, however, the boominess is minor and you want to keep your
loudspeakers, you can make the presentation leaner and tighter by adding low-frequency
absorbers. These acoustic devices soak up low frequencies rather than reflecting them back
into the room. Unlike drapes and bookcases that serve double-duty as conventional domestic furnishings, low-frequency absorbers are purpose-built acoustical devices that may not
aesthetically complement all decors. Nonetheless, adding low-frequency absorption to a listening room often confers a large improvement in the sound quality, specifically bass articulation, dynamics, and midrange clarity. Cleaning up the bass bloat improves the midrange
because removing the excess low-frequency energy no longer obscures or intrudes on the
midrange. As a result, vocals have greater clarity and openness.

     A good example of a low-frequency absorber is the Tube Trap from Acoustic
Sciences Corporation. These cylindrical devices are available in a range of sizes from

8" to 16", with the larger diameter delivering greater absorption at the lowest frequencies. Tube Traps are usually positioned in the corners behind the loudspeakers, which increases their effectiveness (Fig.12-7).

US Patent #4,548,292

THREADED INSERTS FOR STACKING AND HANGING

COMPLETELY SEALED FOR SAFE, DUST-FREE PERFORMANCE

THREE SEPARATE COMPARTMENTS TO CONTROL STANDING WAVES

TREBLE RANGE REFLECTOR

ENTIRE SURFACE IS BASS ABSORPTIVE

ACOUSTICALLY TRANSPARENT FABRIC

STURDY EXOSKELETON CONSTRUCTION FOR STRENGTH AND DURABILITY

SIZES 9, 11, 13, 16, 20" IN DIAMETER AND UP TO 4' IN HEIGHT

©ACOUSTIC SCIENCES CORP.

Fig.12-7 The Tube Trap is extremely effective in controlling low-frequency room problems.
(Courtesy Acoustic Sciences Corporation)

You can often tame boomy bass simply by moving the listening seat forward or backward. The sound energy in a room isn't distributed evenly throughout the room. Instead, stationary patterns of high and low pressure, called standing waves, are set up in the room. Because standing waves occur at low frequencies, a listener in a high-pressure zone (also called a peak) will hear boomy bass. Moving the listening seat forward or backward out of the high-pressure zone will often reduce boomy bass. Conversely, if the sound is excessively thin and lacking bass, you could partially position the listening chair in a high-pressure zone to increase the sense of weight and fullness. Moving the listening chair forward or backward by about two feet is usually enough to affect a change in the bass.

4) Reflective objects near the loudspeaker
Reflective objects near the loudspeakers—equipment racks, windows behind the loudspeakers, subwoofers or furniture between the loudspeakers, even power amplifiers on the floor—can cause poor image focus and lack of depth. The best solution is to remove the offending object. If this isn't possible, move the acoustically reflective object behind the loudspeakers, if possible.

Chapter 12

There's an obvious conflict here for those of you whose audio system is used for both music and home theater; you will likely have a big piece of glass—a television—between your left and right speakers. The solution is to pull the left and right speakers forward of the television by a few inches so that the left, center, and right speakers create an arc. (This technique is described in more detail in Chapter 10).

Acoustical Do's and Don'ts

I've summarized this chapter into a few simple guidelines for improving your listening room. If you just want some practical tips for getting the most out of your system, this is the section to read. More detail on each of these points can be found throughout the chapter.

1) Loudspeaker placement

Just as real estate agents chant "location, location, location" as the three most important things about the desirability of a house, the three most important ways to improve the sound in your room are loudspeaker placement, loudspeaker placement, and loudspeaker placement. Follow the suggestions in this chapter and spend a few hours moving your loudspeakers around and listening. You'll not only end up with better sound, but become more attuned to sonic differences. All acoustic treatments should be built on a foundation of good loudspeaker placement.

2) Avoid untreated parallel surfaces

If you've got bare walls facing each other, you'll have flutter echo. Kill the flutter echo by facing one wall with an absorbent or diffusive material such as drapes, bookcases, or media-storage racks. Dedicated acoustical products work well.

3) Absorb or diffuse side-wall and floor reflections

Bare floors should be covered with carpet between the listening seat and the loudspeakers. Treat the side walls between the loudspeakers and the listening position with an absorbing or diffusing material. Avoid having reflective surfaces, such as bare walls and windows, next to the loudspeakers.

4) Keep reflective objects away from loudspeakers

Equipment racks, power amplifiers, furniture, and other acoustically reflective objects near the loudspeakers will degrade imaging and soundstaging. Move them behind the loudspeakers if possible.

5) Move the listening seat for best low-frequency balance

Standing waves create stationary areas of high and low pressure in the room. Move the listening chair for best balance. Avoid sitting against the rear wall; the sound will be bass-heavy.

6) Break up standing-wave patterns with irregular surfaces or objects

Strategically placed furniture or structures help break up standing waves. Large pieces of furniture behind the listening position diffuse waves reflected from the wall behind the listener.

Digital Room Correction

We've seen in this chapter that the listening room is a crucial link in the music playback chain. Every listening room, no matter how good, radically alters the sound produced by a pair of speakers. The room acts as an equalizer, boosting some frequencies and attenuating others. This change in frequency response can be as much as 30dB in the bass, adding large amounts of coloration to the music. Moreover, the room briefly stores and releases energy at certain frequencies, causing some bass notes to "hang" in the room longer. The result is a thickness and lack of clarity in the bass, along with a reduced sense of dynamic drive.

It's been said that a loudspeaker designer has 100% control over his product's sound above 700Hz, 50% control from 300Hz to 700Hz, and only 20% control below 300Hz. This aphorism reflects the increasing influence of the listening room on sound quality at lower and lower frequencies.

Although the listening room imposes far more severe colorations than any electronic component, we've grown to accept this musical degradation as an inevitable part of music listening. Even custom-built, acoustically designed rooms housing perfectly positioned loudspeakers will introduce significant colorations. All rooms introduce colorations: the better ones simply do less damage.

Attempts to reduce room-induced colorations with analog equalizers have met with limited success. Analog equalizers add their own audible problems, and are limited in resolution. Digital equalizers can have greater resolution than analog ones, but are still a crude tool for dealing with room problems.

The advent of Digital Signal Processing (DSP) provides a powerful new opportunity to remove the effects of the listening room on reproduced music. DSP is the manipulation of audio signals by performing mathematical calculations on the numbers that represent the music. DSP chips can be programmed to perform as filters or equalizers with many hundreds of frequency bands, and with characteristics impossible to realize with analog equalizers. The trick is putting this advanced technology to work to surgically correct room-induced colorations.

Far more sophisticated than a digital equalizer, a DSP room-correction system analyzes the response of your loudspeaker/room combination with high resolution, then creates digital filters to remove the room-induced problems. The room-correction system not only removes the room's effects, but also corrects for speaker colorations. If your speakers have a peak of energy at 2kHz, for example, the room-correction system can remove this coloration. In addition, the room-correction system will make the speaker/room response identical for each speaker. This new technology removes the room's effects on reproduced sound *electronically*, before it is reproduced by the loudspeakers. Put another way, the room-correction system distorts the signal in a way that counteracts the distortion imposed by the listening room. The result is flat response at the listening position.

Note that because room-correction systems operate in the digital domain, analog signals (from LP, FM, portable music player) must be digitized, then converted back to analog. That's why most room-correction systems include an integral analog-to-digital converter. Digital signals, such as from a CD transport, are input directly to the room-correction system, which either converts that digital signal to analog for driving a power amplifier, or outputs a digital signal for conversion to analog with an outboard digital processor. Although room correction can produce profound improvements in bass quality and image focus, some audiophiles will be reluctant to digitize LPs. Analog-to-digital conversion and digital-to-analog conversion are not sonically transparent, although today's converters are much better sounding than those of just a few years ago.

Room-correction systems are rare because the complexity of hardware and software requires a huge expenditure in development time and money. Make no mistake: correcting a room's response is an extraordinarily difficult task. Moreover, a room-correction system is as much a computer as an audio product, requiring specialized design skills not normally found among audio engineers. The first room-correction systems were difficult to set-up and dial in correctly; more modern examples are much easier to use.

A new generation of room-correction systems, however, is much easier to use, doesn't require a computer or computer skills, and is significantly more effective than were the first models. Such a product is shown in Fig.12-8.

Fig.12-8 Today's room-correction systems are much easier to use and more effective than earlier-generation models. (Courtesy Lyngdorf Audio)

My experience with room correction suggests that the technology is extremely effective in removing colorations and restoring clarity to reproduced music. With room correction, the sound has a "lightness," agility, and clarity that are impossible to achieve in any other way. The overall presentation seems to have less midbass boxiness, and extreme low frequencies have greater dynamic impact. In addition, soundstaging improves because identical response from both loudspeakers is essential to precise image focus. (That's why each speaker must be toed-in by exactly the same amount.) Without room correction, image size and placement can change as an instrument moves between musical registers because the response of each speaker/room is different. With identical output from each speaker, image focus becomes tighter and the soundstage is better delineated. Once you get used to the sound of a system with DSP room correction, it's hard to go back.

Accessories and Cable Dressing

There are many setup techniques and accessory products whose individual audible differences may be very small, but that can increase a system's performance when combined. Taking your system to the next level of performance with these products and techniques requires patience, listening skill, and a desire to extract the last bit of performance from your system. There's nothing more satisfying than realizing a significant audible improvement without spending a dime.

All system setup should be based on a foundation of good loudspeaker placement. Once you've made this coarse adjustment, you can work on refinements. As your system's sound improves, you'll become more attuned to small performance variations.

How the cables behind your equipment rack are routed can affect the sound. Keep AC cords away from signal-carrying cables. If they must meet, position them at right angles to each other rather than running them parallel to each other. AC cables radiate 60Hz hum that could be picked up by signal cables. Cables carrying digital data—the output from a CD transport, a DVD player, or other digital source—should be kept away from both AC cables and analog signal-carrying cables. Cables carrying digital data radiate noise that could get into the analog signal.

When playing analog sources, turn off all digital components. Again, digital signals radiate noise. Some A/V controllers automatically shut down unneeded digital circuits when the controller's source-selection switch is set to an analog input.

If you have a power amplifier (or integrated amplifier) in your equipment rack, keep it as far as possible from a phono preamplifier (if your system uses one).

Provide adequate ventilation for heat-producing components. Don't stack a CD player on top of an integrated amplifier, for example. If you must stack components on a single rack shelf, separate them with cones or feet. Overheating may degrade performance, and will certainly shorten product life.

Experiment with isolation feet or cones between your components and the equipment rack. Some are more effective than others, so it's best to try them in your system before purchasing. Most dealers and mail-order suppliers will allow you to trade in feet and cones if they don't improve your system's sound.

Keep interconnects and loudspeaker cables as short as possible. All cable degrades the audio signal; the less of it in your system the better.

Ensure a good contact between the loudspeaker cables and binding posts. Use a nutdriver to get a tight fit, being careful not to over-tighten. Periodically clean the binding post contacts and the speaker cable's spade lugs or banana plugs with a contact cleaner such as Caig DeoxIT. Similarly, clean the RCA jacks on your equipment and the RCA plugs on your interconnects.

Glossary

A/B comparison A back-and-forth listening comparison between two musical presentations, A and B.

absolute polarity A recording with correct absolute polarity played back through a system with correct absolute polarity will produce a positive pressure wave from the loudspeakers in response to a positive pressure wave at the original acoustic event. Incorrect absolute polarity introduces a 180° phase reversal of both channels. Absolute polarity is audible with some instruments and to some listeners.

AC line-conditioner/protector A device that filters noise from the AC powerline and isolates equipment from voltage spikes and surges. Some AC line-conditioners/protectors also protect equipment from lightning strikes. Audio and video equipment is plugged into the AC line-conditioner/protector, and the conditioner is plugged into the wall.

acoustic absorber Any material that absorbs sound, such as carpet, drapes, and thickly upholstered furniture.

acoustic diffuser Any material that scatters sound.

acoustic feedback Sound from a loudspeaker causes a turntable to vibrate; that vibration is converted by the phono cartridge into an electrical signal, which is reproduced by the loudspeaker, which causes the turntable to vibrate even more. This sets up a *feedback loop* in which the vibration feeds on itself, becoming louder and louder. You've probably heard acoustic feedback at an amplified concert as a howl or screech from the PA system.

acoustic panel absorber A device that absorbs low to mid frequencies by diaphragmatic action. That is, sound striking the panel absorber causes the panel to move, converting acoustic energy into a minute amount of heat in the panel.

acoustics The science of sound behavior. Also refers to a room; e.g., "This room has good acoustics."

active subwoofer A speaker designed to reproduce only low frequencies, and which includes an integral power amplifier to drive the speaker.

ADC see "analog-to-digital converter"

AES/EBU interface Professional system for transmitting digital audio. An AES/EBU cable is a balanced line and is terminated with XLR connectors. Also found on some consumer products. Named for the Audio Engineering Society and European Broadcasting Union.

agile Sonic description of bass that can follow quickly changing pitches and dynamics.

aggressive Sonic description of a forward presentation that seems to thrust the music on the listener.

air Sonic description of treble openness, or of space between instruments in the soundstage. Contrast with "dull," "thick."

air-bearing tonearm A tonearm in which the armtube floats on a cushion of air.

air-bearing turntable A turntable in which the platter floats on a cushion of air.

alternate-channel selectivity Tuner specification describing a tuner's ability to reject stations two channels away from the desired station.

ambience Spatial aspects of a film soundtrack that create a sense of size and atmosphere, usually reproduced by the surround speakers.

Glossary

ampere Unit of electrical current, abbreviated A.

analog An analog signal is one in which the varying voltage is an analog of the acoustical waveform; i.e., the voltage varies continuously with the original acoustical waveform. Contrasted with a digital signal, in which binary ones and zeros represent audio or video information.

analog bypass Feature on a digital controller that passes analog input signals to the output without analog-to-digital and digital-to-analog conversions.

analog-to-digital converter A circuit that converts an analog signal to a digital signal. All A/V receivers and A/V controllers have analog-to-digital converters to digitize analog input signals.

analytical Sonic description that describes a component that reveals every nuance in the recording, but in an unpleasant way. An analytical component is rarely musical.

anechoic Literally "without echo."

anechoic chamber An acoustically reflection-free room. An anechoic chamber's walls are covered in highly absorbent material so that no sound is reflected back into the room. Used in loudspeaker testing.

anti-skate adjustment Control on a tonearm that adjusts the amount of force applied to the tonearm to counteract the arm's natural tendency to skate (pull inward).

articulate Sonic description of a component that clearly resolves pitches.

atmosphere see "ambience."

audiophile A person who values sound quality in reproduced music.

A/V Short for audio/video. Identifies a component or system as one that processes video as well as audio signals.

A/V controller Also called an "A/V preamplifier," the A/V controller is a component that lets you control the playback volume and select which source you want to watch. A/V preamplifiers also perform surround decoding.

A/V controller/tuner An A/V controller that includes, in the same chassis, an AM or FM tuner for receiving radio broadcasts.

A/V input An input on an A/V receiver or A/V controller that includes both audio and video jacks.

A/V loop An A/V input and A/V output pair found on all A/V receivers and A/V controllers. Used to connect a component that records as well as plays back audio and video signals.

A/V receiver The central component of a home-theater system; receives signals from source components, selects which signal you watch and listen to, controls the playback volume, performs surround decoding, receives radio broadcasts, and amplifies signals to drive a home-theater loudspeaker system. Also called a "surround receiver."

baffle The front surface of a loudspeaker on which the drivers are mounted.

balanced connection A method of connecting audio components with three conductors in a single cable. One conductor carries the audio signal, a second conductor carries that audio signal with inverted polarity, and the third conductor is the ground.

banana jack A small tubular connector found on loudspeakers and power amplifiers for connecting speaker cables terminated with banana plugs.

banana plug A common speaker-cable termination that fits into a banana jack.

bandwidth The range of frequencies that a device can process or pass. In reference to an electrical or acoustic device, bandwidth is the range of frequencies between the −3dB points.

bass Sounds in the low audio range, generally frequencies below 500Hz.

bass extension A measure of how deeply an audio system or loudspeaker will reproduce bass. For example, a small subwoofer may have bass extension to 40Hz. A large subwoofer may have bass extension to 16Hz.

bass management A combination of controls and circuits in an A/V receiver, A/V controller, or multichannel digital audio player (DVD-A and SACD) that determines which speakers receive bass signals.

bass reflex A speaker design with a hole or slot in the cabinet that allows sound inside the cabinet to emerge into the listening room. Bass reflex speakers have deeper bass extension than speakers with sealed cabinets, but that bass is generally less tightly controlled. Also called "vented" or "ported." Contrast with "infinite baffle" or "air suspension."

bi-amping Using two power amplifiers to drive one loudspeaker. One amplifier typically drives the woofer, the second drives the midrange and tweeter.

binding post A connection on power amplifiers and loudspeakers for attaching loudspeaker cables.

bipolar speaker A speaker that produces sound equally from the front and the back. Unlike the dipolar speaker, the bipolar's front and rear soundwaves are in-phase with each other.

bit rate The number of bits per second stored or transmitted by a digital audio or digital video signal. For example, the bit rate of compact disc is 705,600 bits per second per channel. Dolby Digital has a bit rate of 384kbs (384,000 bits per second) for 5.1 channels. MP3 has a bit rate as low as 32kbs per channel. DVD-Audio has a bit rate as high as 4.608 million bits per second per channel. Higher bit rates generally translate to better audio quality. Also called "data rate."

bi-wiring Technique of running two cables to each loudspeaker. One cable is connected to the loudspeaker's woofer input terminals, one to the tweeter's input terminals. Bi-wiring is possible only with loudspeakers with two pairs of input terminals.

Blu-ray Disc Disc format for storing high-definition video as well as high-resolution digital audio.

boomy Excessive bass, particularly over a wide band of frequencies.

break-in Initial period of use of a new audio component, during which time the component's sound improves.

bridging Amplifier-to-loudspeaker connection method that converts a stereo amplifier into a monoblock power amplifier. One amplifier channel amplifies the positive half of the waveform, the other channel amplifies the negative half. The loudspeaker is connected as the "bridge" between the two amplifier channels.

brightness In audio, an excessive amount of treble that adds shrillness to the sound.

brittle A midrange or treble character that makes instrumental timbres sound harsh. Contrast with "liquid."

build-quality The quality of electronic parts, chassis, and construction techniques of an audio or video component.

Glossary

calibration The act of fine-tuning an audio or video component for correct performance. In an audio system, calibration includes setting the individual channel levels. In video, calibration means setting a video display device to display the correct color, brightness, tint, contrast, and other parameters.

cantilever Thin tube protruding from a phono cartridge that holds the stylus.

capacitor Electronic component that stores a charge of electrons. Reservoir capacitors are used for energy storage in power amplifiers; filter capacitors filter traces of AC from DC power supplies; coupling capacitors connect one amplifier stage to another by blocking DC and allowing the AC audio signal to pass.

capture ratio Tuner specification: the difference in dB required between the strengths of two stations needed before a tuner can lock to the stronger station and reject the weaker one. The lower the capture ratio, the better the tuner.

cartridge demagnetizer Device that removes stray magnetic fields from metal parts inside a phono cartridge.

center channel In a multichannel audio system, the audio channel that carries information that will be reproduced by a speaker placed in the center of the viewing room between the left and right speakers. The center channel carries nearly all of a film's dialogue.

center-channel speaker The speaker in a home-theater system located on top of, beneath, or behind the visual image; reproduces center-channel information such as dialogue and other sounds associated with onscreen action.

channel balance The relative levels or volumes of the left and right channels in an audio system or individual component.

channel separation A measure of how well sounds in one channel are isolated from the other channels.

chuffing Sound created by the port of a bass-reflex loudspeaker when reproducing low bass at high levels. Caused by large movement of air in the port.

Class-A Mode of amplifier operation in which a transistor or tube amplifies the entire audio signal.

Class-A/B Mode of amplifier operation in which the output stage operates in Class-A at low output power, then switches to Class-B at higher output power.

Class-B Mode of amplifier operation in which one tube or transistor amplifies the positive half of an audio signal, and a second tube or transistor amplifies the negative half.

Class-D Mode of amplifier operation in which the output transistors are switched fully on or fully off.

clipping An amplifier that is asked to deliver more power than it is capable of will flatten the tops and bottoms of the audio waveform, making the peaks appear to be clipped off. Clipping introduces a large amount of distortion, audible as a crunching sound on musical peaks.

coaxial cable A cable in which an inner conductor is surrounded by a braided conductor that acts as a shield.

coaxial digital output A jack found on most digital source products (disc players, satellite receivers, music servers) that provides a digital audio signal on an RCA jack for connection to another component through a coaxial digital interconnect.

coherence The impression that the music is an integrated whole, rather than made up of separate parts.

202

coloration A change in sound introduced by a component in an audio system. A "colored" loudspeaker doesn't accurately reproduce the signal fed to it. A speaker with coloration may have too much bass and not enough treble, for example.

compliance In phono cartridges, a number expressing the cantilever's stiffness. Specifically, compliance is the length of cantilever movement when a force of 10^{-6} dynes is applied, expressed in millionths of a centimeter. High compliance is any value above 20. Low compliance is any value below 10.

cone The paper, plastic, or metal diaphragm of a loudspeaker that moves back and forth to create sound.

contact cleaner Fluid for removing oxide and dirt from audio jacks and plugs.

controller see A/V "controller"

critical listening The art of evaluating audio equipment quality by careful analytical listening for specific sonic flaws.

crossover A circuit that splits up the frequency spectrum into two or more parts. Crossovers are found in virtually all dynamic loudspeakers, and in A/V receivers and A/V controllers.

crossover frequency The frequency at which the audio spectrum is split. A subwoofer with a crossover frequency of 80Hz filters all information above 80Hz out of the signal driving the subwoofer, and all information below 80Hz out of the signal driving the main speakers.

crossover slope Describes the steepness of a crossover filter. Expressed as "xdB/octave." For example, a subwoofer with a crossover frequency of 80Hz and a slope of 6dB/octave would allow audio frequencies at 160Hz (an octave above 80Hz) into the subwoofer, but signals at 160Hz would be reduced in amplitude by 6dB. A slope of 12dB/octave would also allow 160Hz into the subwoofer, but the amplitude would be reduced by 12dB. The most common crossover slopes are 12dB/octave, 18dB/octave, and 24dB/octave. Crossover slopes are also referred to as "first-order" (6dB/octave), "second-order" (12dB/octave), "third-order" (18dB/octave), and "fourth-order" (24dB/octave). The "steeper" slopes (such as 24dB/octave) split the frequency spectrum more sharply and produce less overlap between the two frequency bands.

crosstalk see "channel separation"

current The flow of electrons in a conductor. For example, a power amplifier "pushes" electrical current through speaker cables and the voice coils in a loudspeaker to make them move back and forth.

DAC see "digital-to-analog converter"

damping factor A number that expresses a power amplifier's ability to control woofer motion. Related to the amplifier's output impedance.

data rate see "bit rate"

dB see "decibel"

DC (Direct Current) Flow of electrons that remains steady rather than fluctuating. Contrasted with alternating current (AC).

decibel The standard unit for expressing relative power or amplitude levels of voltage, electrical power, acoustical power, or sound-pressure level. Abbreviated dB.

depth The impression of instruments, voices, or sounds existing behind one another in three dimensions, as in "soundstage depth."

Glossary

detail Low-level components of the musical presentation, such as the fine inner structure of an instrument's timbre.

diaphragm The surface of a loudspeaker driver that moves, creating sound.

diffraction The bending of soundwaves as they pass around an object. Also: a re-radiation of sound caused by discontinuities in surfaces near the radiating device, such as the bolts that secure drivers to a speaker cabinet.

diffusion Scattering of sound. Diffusion reduces the sense of direction of sounds, which benefits sound produced by surround loudspeakers.

digital Calculation or representation by discrete units. For example, digital audio and digital video can be represented by a series of binary ones and zeros.

digital audio server Source component that stores music on hard-disc drives and can be programmed to play music by genre or other selection criteria.

digital loudspeaker Loudspeaker incorporating a digital crossover and power amplifiers. A digital loudspeaker takes in a digital bitstream, spits up the frequency spectrum with digital signal processing, converts each of those signals to analog, and amplifies them separately. The individual power amplifiers then power each of the loudspeaker's drive units.

Digital Signal Processing (DSP) Manipulation of audio or video signals by performing mathematical functions on the digitally encoded signal.

digital-to-analog converter A chip that converts digital audio signals to analog audio signals. Digital source components (disc players, satellite receivers, music servers) all contain digital-to-analog converters (but CD transports do not). Also: A stand-alone component in an audio system that converts digital audio signals to analog signals. Also called a digital processor or DAC.

Digital Transmission Content Protection (DTCP) Copy-protection system that allows content owners to control how the content may be copied. Also known as "5C" after the five companies that developed the standard.

dip A reduction in energy over a band of frequencies. Contrast with peak.

dipolar speaker A loudspeaker that produces sound from the rear as well as from the front, with the front and rear sounds out-of-phase with each other.

Direct Stream Digital (DSD) Method of digitally encoding music with a very fast sampling rate (2.8224MHz), but with only 1-bit quantization. Developed by Sony and Philips for the Super Audio CD (SACD) format.

discrete Separate. A circuit using separate transistors rather than an integrated circuit. A discrete digital surround-sound format contains 5.1 channels of audio information that are completely separate from each other; contrasted with a matrixed surround format such as Dolby Surround, which mixes the channels together for transmission or storage.

dispersion The directional pattern over which a loudspeaker distributes its sound.

Dolby Digital A 5.1-channel discrete digital surround-sound format used in movie theaters and consumer formats. The surround format required on DVD disc.

Dolby Digital Plus An extension of Dolby Digital that employs better encoding algorithms as well as a much higher bit rate to deliver superior sound. Developed for HD DVD and Blu-ray Disc.

Dolby Digital EX Surround format that matrix-encodes a third surround channel in the existing left and right surround channels of a Dolby Digital signal. This third surround channel (called "surround back") drives a loudspeaker or loudspeakers located directly behind the listening position for greater spatial realism. Called THX EX in THX-certified products.

Dolby Pro Logic A type of Dolby Surround decoder with improved performance over standard Dolby Surround decoding. Specifically, Pro Logic decoding provides greater channel separation and a center-speaker output. A Dolby Pro Logic decoder takes in a 2-channel, Dolby Surround-encoded audio signal and splits that signal up into left, center, right, and surround channels.

Dolby Pro Logic II Pro Logic surround decoder with improved performance and added features compared with older Pro Logic circuits.

Dolby Pro Logic IIx Pro Logic surround decoder that generates the surround-back channels from 2-channel or 5.1-channel sources.

Dolby Surround An encoding format that combines four channels (left, center, right, surround) into two channels for transmission or storage. On playback, a Dolby Pro-Logic decoder separates the two channels back into four channels.

Dolby TrueHD Surround-sound format that delivers high-resolution multichannel digital audio losslessly; that is, with perfect bit-for-bit accuracy to the source. Developed as an option for HD DVD and Blu-ray Disc.

driver The actual speaker units inside a loudspeaker cabinet.

DSD see "Direct Stream Digital"

DSP see "Digital Signal Processing"

DSP room correction Technique of removing room-induced frequency-response peaks and dips with digital signal processing.

DTCP see "Digital Transmission Content Protection"

DTS A discrete digital surround-sound format used in movie theaters and some home-theater systems. A better-sounding alternative to Dolby Digital. Also called DTS Digital Surround. The DTS format encompasses film soundtracks and music releases.

DTS ES Discrete Surround-sound format that delivers discrete 6.1-channel sound. The additional channel (compared with 5.1-channel sound) drives a loudspeaker or loudspeakers located directly behind the listening position.

DTS ES Matrix Surround-sound format that matrix-encodes a third surround channel into the existing left and right surround channels in a DTS signal.

DTS-HD Surround-sound format developed for HD DVD and Blu-ray Disc that delivers superior sound quality. Has a much higher bit rate than DTS.

DTS-HD Master Audio Surround-sound format that can deliver high-resolution multichannel digital audio losslessly; that is, with perfect bit-for-bit accuracy to the source.

DTS Neo:6 Decoding circuit developed by DTS for creating 5.1-channel audio from 2-channel sources. Comes in two flavors: DTS Neo:6 Music and DTS Neo:6 Cinema.

DualDisc Disc format that contains a DVD-Video disc on one side and DVD-Audio on the other side.

dull Lacking treble energy.

DVD-Audio Subset of the DVD standard that provides for high-resolution, multichannel digital audio on DVD.

Glossary

dynamic range In audio, the difference in volume between loud and soft. In video, the difference in light level between black and white (also called contrast).

electrostatic loudspeaker Loudspeaker in which a thin diaphragm is moved back and forth by electrostatic forces. Contrasted with a dynamic loudspeaker, in which electromagnetic forces move the diaphragm back and forth.

equalization In tape or LP record playback, a treble cut to counteract a treble boost applied during recording. Also describes modification of tonal balance by employing an equalizer.

equalizer A circuit that changes the tonal balance of an audio program. Bass and treble controls are a form of equalizer.

extension How high or low in frequency an audio component can reproduce sound.

filter Electronic circuit that selectively removes or reduces the amplitude of certain frequencies.

5.1 channel sound The standard number of channels for encoding film soundtracks. The five channels are left, center, right, surround left, and surround right. The ".1" channel is a 100Hz-bandwidth channel reserved for high-impact bass effects, called "low-frequency effects," or LFE.

flat A speaker that accurately reproduces the signal fed to it is called "flat" because that is the shape of its frequency-response curve. Flat also describes a soundstage lacking in depth.

floorstanding speaker A speaker that sits on the floor rather than on a stand.

flutter echo Back-and-forth acoustic reflections in a room between pairs of surfaces. Think of a pair of facing mirrors, each reflecting light into the other. Flutter echo can be heard as a "pinging" sound after a handclap. Caused by untreated parallel surfaces.

forward A description of a sonic presentation in which sounds seem to be projected forward toward the listener.

frequency Number of repetitions of a cycle. Measured in Hertz (Hz), or cycles, per second. An audio signal with a frequency of 1000Hz (1kHz) undergoes 1000 cycles of a sinewave per second.

frequency response A graphical representation showing a device's relative amplitude as a function of frequency.

full-range speaker A speaker that reproduces bass as well as midrange and treble frequencies.

geometry In cables, the physical arrangement of the conductors and dielectric.

grainy Sonic description of a roughness to instrumental or vocal timbres.

harmonic distortion The production of spurious frequencies at multiples of the original frequency. A circuit amplifying a 1kHz sinewave will create frequencies at 2kHz (second harmonic), 3kHz (third harmonic), and so forth.

HDCD see "High Definition Compatible Digital"

HD DVD Disc format designed as the high-definition replacement for DVD.

heatsink Large metal device that draws heat away from the interior of an electronic device and dissipates that heat in air. The fins protruding from the sides of power amplifiers are heatsinks.

HDMI High-Definition Multimedia Interface. Interface format for transmitting high-definition video as well as high-resolution multichannel digital audio, all in the same cable.

Hertz (Hz) The unit of frequency; the number of cycles per second. Kilohertz (kHz) is thousands of cycles per second.

High Definition Compatible Digital (HDCD) Process for improving the sound quality of 16-bit/44.1kHz digital audio on compact disc. An HDCD-encoded disc will play on any CD player, but sounds best when played on a CD player or digital processor equipped with an HDCD decoder.

high-density layer The information layer in a hybrid Super Audio CD that contains high-resolution digital audio.

high-pass filter A circuit that allows high frequencies to pass but blocks low frequencies. Also called a "low-cut filter." High-pass filters are often found in A/V receivers and A/V preamplifiers to keep bass out of the front speakers when you're using a subwoofer.

high-resolution digital audio Generally regarded as digital audio with a sampling rate greater than 48kHz and a word length longer than 16 bits.

home theater The combination of high-quality sound and video in the home.

Home THX A set of patents, technologies, and playback standards for reproducing film soundtracks in the home. THX doesn't compete with surround formats such as Dolby Digital or DTS, but instead builds on them.

hybrid An audio component that combines more than one technology, such as tubes and transistors in the same amplifier, or dynamic and ribbon drivers in the same loudspeaker.

hybrid disc SACD-based disc that is compatible with CD players as well as SACD players. Has two information layers; one layer contains CD-quality audio, the second contains high-resolution audio.

Hz see "Hertz"

IC Integrated circuit. Some products use ICs for processing and amplifying audio signals; higher-quality units use discrete transistors instead.

i.LINK Sony tradename for FireWire (also known as IEEE1394), a high-speed digital interface.

imaging The impression of hearing, in reproduced music, instruments and voices as objects in space.

immediate An immediate musical presentation is somewhat vivid, lively, and forward. Contrast with "laid-back."

impedance Resistance to the flow of AC electrical current. An impedance is a combination of resistance, inductive reactance, and capacitive reactance.

infinite baffle A sealed loudspeaker cabinet. The cabinet wraps around the drive units, mimicking a baffle of infinite size. Also called air suspension or acoustic suspension. Contrast with a reflex, or ported, loudspeaker.

input impedance The resistance to current flow presented by a circuit or component to the circuit or component driving it. Impedance is a combination of resistance, capacitive reactance, and inductive reactance.

integrated amplifier Audio product combining a preamplifier and power amplifier in one chassis.

interconnect A cable that carries line-level audio signals (audio interconnect), composite video signals (video interconnect), or S/PDIF-encoded digital audio (digital interconnect.).

Glossary

in-wall speaker A speaker that can be mounted inside the cavity cut into a wall.

jitter Timing variations in the clock that synchronizes events in a digital audio system. The clock could be in an analog-to-digital converter that controls when each audio sample is taken. Of more interest to audiophiles is clock jitter in digital audio reproduction; the clock controls the timing of the reconstruction of digital audio samples into an analog signal. Jitter degrades musical fidelity.

kbs Kilo-bits per second. Thousands of bits per second; a measure of bit rate.

kHz see "kilohertz"

kilohertz Thousands of Hertz. A frequency of 1000Hz is 1kHz, for example.

lean Sonic description of a musical presentation lacking midbass. Synonyms are "thin," "lightweight," and "underdamped" (to describe loudspeakers). Contrast with "weighty," "full," and "heavy."

level matching Technique of ensuring that two musical presentations are reproduced at exactly the same volume so that more accurate judgments of the audio quality can be reached.

LFE see "Low Frequency Effects"

line level An audio signal with an amplitude of approximately 1V to 2V. Audio components interface at line level through interconnects. Contrasted with "speaker level," the much more powerful signal that drives speakers.

line-source loudspeaker Loudspeaker with a tall and narrow dispersion pattern. A tall ribbon driver is naturally a line source, as is a vertical array of point-source drivers. Contrasted with a point-source loudspeaker, which has a shorter, broader dispersion pattern.

liquid Sonic description of a musical presentation lacking shrillness. Usually applied to the midrange, liquidity implies correct reproduction of musical timbre.

localization The ability to detect the directionality of sounds.

lossless coding Method of reducing the bit rate of a digital audio signal while maintaining perfect bit-for-bit accuracy with the source data. Compare with "lossy" coding.

loudspeaker A device that converts an electrical signal into sound. The loudspeaker is the last component of the playback chain.

low-cut filter A circuit that removes bass frequencies from an audio signal. Also called a "high-pass filter."

low-frequency cutoff The point at which a loudspeaker's output drops in the bass by 3dB.

Low Frequency Effects (LFE) A separate channel in the multichannel film-sound-track formats reserved for low bass effects, such as explosions. The LFE channel is the ".1" channel in 5.1-channel and 7.1-channel sound.

low-pass filter A circuit that removes midrange and treble frequencies from an audio signal. Also called a "high-cut filter."

Mbs Mega (million) bits per second. A unit of measure for expressing bit rates. MPEG-2 video encoding has a variable bit rate that averages 3.5Mbs.

MDF see "Medium Density Fiberboard"

Medium Density Fiberboard (MDF) Composite wood material from which most loudspeaker cabinets are made.

Meridian Lossless Packing (MLP) Data compression system used in DVD-Audio that reduces the bit rate with no loss in quality. The decompressed bitstream is bit-for-bit identical to the original bitstream. Contrasted with "lossy" compression systems that degrade fidelity.

midrange Audio frequencies in the middle of the audible spectrum, such as the human voice. Generally the range of frequencies from about 300Hz to 2kHz. Also: a driver in a loudspeaker that reproduces the range of frequencies in the middle of the audible spectrum.

millisecond One one-thousandth of a second.

minimonitor A small, stand-mounted loudspeaker.

MLP see "Meridian Lossless Packing"

monoblock A power amplifier with only one channel.

moving-coil cartridge Transducer that converts stylus motion in a record groove to an electrical signal. Tiny coils attached to the cantilever are moved back and forth in a fixed magnetic field, inducing current flow through the coils.

moving-magnet cartridge Transducer that converts stylus motion in a record groove to an electrical signal. Tiny magnets attached to the cantilever are moved back and forth between fixed coils of wire, inducing current flow through the coils.

MP3 Low-bit-rate coding scheme used in portable audio players and Internet downloads.

multichannel power amplifier A power amplifier with more than two channels, usually five or seven.

multichannel preamplifier A preamplifier with more than two channels, usually six.

multichannel sound Sound reproduction via more than two channels feeding more than two loudspeakers.

multi-room A feature on some A/V products that lets you listen to two different sources in two different rooms.

octave The interval between two frequencies with a ratio of 2:1. The bottom octave in audio is 20Hz-40Hz; the top octave is 10kHz-20kHz.

off-axis response A loudspeaker's frequency response measured at the loudspeaker's sides. Contrast with on-axis, the loudspeaker's response directly in front of the baffle.

ohm The unit of resistance to electrical current flow.

on-axis response A loudspeaker's frequency response measured directly in front of the baffle.

one-note bass Sonic description of bass that seems to have just one pitch. Caused by excessive output over a narrow frequency band in loudspeakers. A "boom truck" is one-note bass taken to an extreme.

on-wall speaker A speaker with a shallow depth that can be mounted on a wall, usually next to a flat-panel television.

output stage The last amplifier circuit in an audio component. The output stage in a CD player is an amplifier that drives the preamplifier. In power amplifiers, the output stage delivers current to the loudspeakers.

Glossary

output transformer Transformer in tubed amplifiers that couples the output stage to the loudspeaker. Output transformers are required in tubed amplifiers to change the amplifier's high output impedance to a lower value that can better drive loudspeakers. The output transformer also blocks DC from appearing at the amplifier's output terminals.

passive radiator Diaphragm in some loudspeakers that isn't connected electrically, but is moved by air pressure inside the cabinet created by the woofer's motion. Also called an auxiliary bass radiator (ABR), the passive radiator covers what would have been the port in a reflex-loaded loudspeaker.

passive subwoofer A speaker for reproducing bass frequencies that must be powered by a separate power amplifier. Contrasted with "active" or "powered" subwoofers, which contain an integral amplifier.

PCM see "Pulse Code Modulation"

peak A short-term, high-level audio signal. Also: an excess of energy over a narrow frequency band (contrasted with "dip").

peaky Sonic description of a sound with excessive energy over a narrow frequency band.

phantom image The creation of an apparent sound source between two loudspeakers.

phase In a periodic wave, the fraction of a period that has elapsed. Describes the time relationship between two signals.

phase adjustment A control provided on some subwoofers that lets you delay the sound of the subwoofer slightly so that the subwoofer's output is in-phase (has the same time relationship) with the front speakers.

pitch definition The ability to distinguish pitch in reproduced music, particularly in the bass. Some products (especially loudspeakers) obscure the individual pitches of notes.

pivoted tonearm A tonearm in which the cartridge and armtube traverse the record in an arc while maintaining a fixed pivot point. Contrasted with tangential-tracking tonearms.

planar loudspeaker Loudspeaker in which the driver or drivers are mounted in an open panel.

planar magnetic Type of driver in which conductors carrying the audio signal are bonded to a diaphragm. A subset of the ribbon driver, planar magnetic drivers are also called quasi-ribbons.

plinth The flat deck of a turntable beneath the platter.

point-source loudspeaker Loudspeaker that emits sound from a point in space. Contrast with line-source loudspeaker.

port Opening in a loudspeaker cabinet that channels bass from inside the enclosure to outside the enclosure. Also called a "vent."

ported loudspeaker see "reflex-loaded loudspeaker"

port noise Noise generated by large air flows in a reflex-loaded loudspeaker's port. Also called "chuffing."

power amplifier An audio component that boosts a line-level signal to a powerful signal that can drive loudspeakers.

power handling A measure of how much amplifier power, in watts, a speaker can take before it is damaged.

power output A measure of a power amplifier's ability, in watts, to deliver electrical voltage and current to a speaker.

power supply Circuitry found in every audio component that converts 60Hz alternating current from the wall outlet into direct current that supplies the audio circuitry.

power transformer Device in a power supply that reduces the incoming voltage from 120V to a lower value.

preamplifier Component that receives signals from source components, selects a source for listening, controls the volume, and drives the power amplifier. Literally means "before the amplifier."

presence The sense that an instrument or voice is actually in the listening room.

presence region Band of frequencies in the midrange that contributes to presence.

Pulse Code Modulation (PCM) A method of representing an audio signal as a series of digital samples.

punchy Sonic presentation having dynamic impact, particularly in the bass.

radiation pattern The way in which a speaker disperses sound.

RCA jack A connector found on audio and video products. RCA jacks can carry line-level analog signals as well as digital signals (linear PCM from CD sources in the S/PDIF format, and Dolby Digital or DTS from DVD players, for examples).

Red Book Name for the compact disc specification, derived from the document's red cover. A "Red Book" disc is a conventional CD.

Red Book layer The information layer in a Super Audio CD that contains conventional 16-bit/44.1kHz digital audio.

reflex-loaded loudspeaker see "bass reflex"

resolution The quality of an audio component that reveals low-level musical information.

resonance Vibration of an object or air that is disproportionate in amplitude to the stimulus. A bell ringing at a certain frequency is an example of resonance; the pitch is the bell's resonant frequency.

reverberation Dense acoustical reflections in an acoustic space that become lower in amplitude and more closely spaced over time. The sound in a room after the sound source has stopped producing sound.

reverberation time The time it takes sound in a room to decrease in amplitude by 60dB. Symbol: RT_{60}.

RIAA Recording Industry Association of America

RIAA accuracy Flatness of the RIAA phono equalization circuit in a phono pre-amplifier.

RIAA equalization A treble boost and bass cut applied to the audio signal when a record is cut; a reciprocal treble cut and bass boost on playback restores flat response. RIAA equalization increases the playing time of an LP (because bass takes up the most room in the record groove) and decreases noise (because the treble cut on playback also reduces record surface noise).

ribbon loudspeaker Loudspeaker in which the diaphragm is electrically conductive and carries the audio signal. Usually made from a long, thin strip of aluminum.

rolled off Sonic description of reduced energy at the frequency extremes (bass or treble). A loudspeaker whose treble is rolled off sounds dull.

211

Glossary

rolloff Reduction in energy at the frequency extremes, or the effect of a filter; e.g. the crossover produces a 12dB per octave rolloff above 2kHz.

room correction see "DSP room correction"

room gain Increase in bass level when a loudspeaker plays in a room compared with the loudspeaker's bass level in an anechoic chamber. The room's walls increase the amount of bass heard; the closer to the walls the loudspeaker is placed, the greater the room gain.

room resonance modes Excessive acoustical energy at certain frequencies when the air in a room is excited by the sound of a loudspeaker.

rumble Low-frequency noise associated with LP playback.

SACD see "Super Audio CD"

sampling The process of converting an analog audio signal into digital form by taking periodic "snapshots" of the audio signal at some regular interval. Each snapshot (sample) is assigned a number that represents the analog signal's amplitude at the moment the sample was taken.

sampling frequency The rate at which samples are taken when converting analog audio to digital audio. Expressed in samples per second, or, more commonly, in Hertz; i.e., the CD format's sampling frequency is 44.1kHz.

sampling-rate converter Circuit that changes the sampling frequency of a digital audio signal.

satellite speaker A small speaker with limited bass output designed to be used with a subwoofer.

selectivity Tuner specification describing the tuner's ability to reject unwanted stations. Good selectivity is important to those who live in cities, where stations are closely spaced on the broadcast spectrum.

sensitivity 1) A measure of how much sound a speaker produces for a given amount of input power. Speaker sensitivity is measured by driving a speaker with 1W of power and measuring the sound-pressure level from a distance of 1 meter. 2) A measure of an FM tuner's ability to pull in weak stations. Tuner sensitivity is important if you live a long distance from FM transmitters.

servo-driven woofer A woofer in which an accelerometer (a device that converts motion to an electrical signal) attached to the voice coil sends to an amplifier information about its position and motion. The amplifier then applies a correction signal so that the woofer's motion matches the characteristics of the audio signal.

shielded loudspeaker A loudspeaker lined with metal to contain magnetic energy within the speaker. Shielded loudspeakers are used in home theater because the speakers' magnetic energy can distort a CRT video monitor's picture. Shielding is not necessary with flat-panel televisions.

sibilance *s* and *sh* sounds in spoken word or singing.

signal-to-noise ratio Numerical value expressing in decibels the difference in level between an audio component's noise floor and some reference signal level.

single-ended amplifier Amplifier in which both half-cycles of the audio waveform are amplified by the output tube or transistors. Contrasted with "push-pull amplifier."

single-presentation listening Evaluating audio components by listening to just that component rather than in comparison with other components.

six-channel input An AVR, controller, or multichannel preamplifier input comprising six discrete jacks that will accept the six discrete analog outputs from a multichannel SACD or DVD-Audio player.

skating Force generated in LP playback that pulls the tonearm toward the record center.

skin effect Phenomenon in cables in which high frequencies travel along the conductor's surface and low frequencies travel through the conductor's center.

smooth Sonic description of a presentation lacking peaks and dips in the frequency response.

sound-pressure level (SPL) A measure of loudness expressed in decibels (dB).

soundstage The impression of soundspace existing in three dimensions in front of or around the listener.

source components Components that provide audio or video signals to the rest of the system. Disc players, turntables, FM tuners, music servers, and satellite receivers are all source components;

source switching Function performed by preamplifier or integrated amplifier that selects which source component's signals are fed to the speakers.

spade lug A speaker termination with a flat area that fits around a binding post.

S/PDIF interface Standardized method of transmitting digital audio from one component to another. Stands for Sony/Philips Digital Interface Format.

speaker see "loudspeaker"

SPL see "sound-pressure level"

SPL meter A device for measuring the Sound-Pressure Level created by an audio source.

sprung turntable A turntable in which the platter and armboard are mounted on a sub-chassis that floats within the base on springs. Contrasted with unsprung turntables.

standing wave Stationary area of high and low acoustical pressure in a room caused by interaction of the sound with the room's boundaries.

stator The element in an electrostatic loudspeaker driver that remains stationary.

step-up transformer Transformer with a higher output voltage than the input voltage. Sometimes used between a moving-coil cartridge and a moving-magnet phono input.

stylus Tiny wedge protruding from a phono cartridge's cantilever that rests in the record groove and moves back and forth in response to modulations in the groove.

sub-chassis A component of some sprung turntables in which a platform (the sub-chassis) is suspended from the plinth, and upon which the platter and tonearm are mounted.

subwoofer A speaker designed to reproduce low bass frequencies.

Super Audio CD (SACD) Sony/Philips format for storing CD-quality audio as well as high-resolution digital audio on a CD-sized disc. "Hybrid" SACD discs can be played on conventional CD players as well as SACD players.

surround-back speakers Speakers located directly behind the listening position to reproduce the surround-back channels through a 7.1-channel loudspeaker array.

surround decoder A circuit or component that converts a surround-encoded audio signal into multiple audio signals that can be amplified. All A/V controllers incorporate surround decoding to convert the Dolby Digital or DTS signals from DVD, satellite, or over-the-air HDTV broadcasts to discrete analog signals.

Glossary

surround mode A setting on A/V receivers and A/V controllers that determines what surround decoding or signal processing is performed on the audio signal.

surround sound An audio recording and playback format that uses more than two channels, and is reproduced with more than two loudspeakers, some of which are located behind the listener.

surround speakers Speakers located beside or behind the listener that reproduce the surround channel of surround-sound-encoded audio programs.

sweet spot Point in the listening room where the sound is the best.

system matching The art of combining components to create the most musical system for a given budget.

terminations The fittings on the end of a cable: RCA plugs, spade lugs, banana plugs, etc.

THD see "Total Harmonic Distortion"

theater pass-through Feature on some controllers that sets the controller to some fixed gain setting (usually unity gain) independent of the volume control setting. A required feature if your system uses both a 2-channel preamplifier and an A/V controller.

three-way speaker A loudspeaker that divides the frequency spectrum into three parts (bass, midrange, treble) for reproduction through three or more drivers.

THX Set of patents, technologies, and technical/acoustic performance criteria for film-sound reproduction in movie theaters. (see also "Home THX")

THX-certified An A/V product that correctly implements the THX technologies and meets stringent technical performance criteria for film-sound reproduction.

timbre The physical quality of a sound.

toe-in Angling loudspeakers so that they point directly toward the listening position rather than straight ahead.

tonal balance Relative levels of bass, midrange, and treble in an audio component or musical presentation.

TosLink cable An optical cable for carrying digital audio, either linear PCM from a CD transport or a multichannel format (Dolby Digital or DTS) from a DVD player or other digital source to an A/V receiver or A/V controller.

total harmonic distortion (THD) Specification stating the amount of harmonic distortion in an audio component. Called "total" because it is the sum of all the individual harmonic-distortion components created by the component.

tracking error In LP playback, a difference in the stylus/groove relationship between the cutting stylus and playback stylus.

transducer Any device that converts energy from one form to another. Microphones, loudspeakers, and phono cartridges are transducers.

transient A short-lived sound, often at high level. The sound of a snare drum is an example of a musical transient.

transient response The ability of a component to accurately reproduce transient musical events.

transistor Device made from solid semi-conductor material that can amplify audio signals.

transparent Sonic description of a component or system that has very low levels of coloration. A soundstage in which the acoustic space sounds clear rather than veiled.

214

treble High audio frequencies, generally the range from 3kHz-20kHz.

triode The simplest vacuum tube, employing just three elements: the cathode, plate, and control grid.

tube see "vacuum tube"

tweeter A speaker driver designed to reproduce treble signals.

two-way speaker A loudspeaker that splits the frequency spectrum into two parts, bass and treble, for reproduction by two or more drivers.

unbalanced connection Connection method in which the audio signal is carried on two conductors, called signal and ground. Contrasted with balanced connection, in which the audio signal is carried on three conductors.

universal disc player Source component that plays CD, DVD-Video, DVD-Audio, and SACD.

unsprung turntable Turntable in which the platter and tonearm are connected directly to the turntable base rather than suspended on a sub-chassis. Contrast with "sprung turntable."

usable sensitivity Tuner specification that states the voltage across the antenna required to produce an audio signal with a signal-to-noise ratio of 30dB. Contrast with the more stringent "quieting sensitivity."

user interface The controls and displays on a product and their logic and ease of use.

vacuum tube Device for amplifying audio in which the active elements are enclosed in a glass envelope devoid of air.

veiled Impression of a haze or veil between you and the musical presentation. Contrast with "transparent."

vertical tracking angle (VTA) The angle at which the stylus sits in a record's groove. Adjusted by moving the tonearm pivot point up or down.

vertical tracking force (VTF) Pressure applied by gravity to the stylus in a record groove.

vivid Musical presentation in which every sound is clearly audible.

voice coil Coil of wire inside a loudspeaker driver through which current from the power amplifier flows.

volt Unit of electromotive force. The difference in potential required to make one Ampere of current flow through one ohm of resistance. See also "voltage."

voltage Analogous to electrical pressure. Voltage exists between two points when one point has an excess of electrons in relation to the other point. A battery is a good example: the negative terminal has an excess of electrons in relation to the positive terminal. If you connect a piece of wire between a battery's positive and negative terminals, voltage pushes current through the wire. One volt across 1 ohm of resistance produces a current of 1 Ampere.

VTA see "Vertical Tracking Angle"

VTF see "Vertical Tracking Force"

watt The unit of electrical power, defined as the power dissipated by 1 Ampere of current flowing through 1 ohm of resistance.

wavelength The distance between successive cycles of a sinewave or other periodic motion.

woofer Driver in a loudspeaker system that reproduces bass.

XLR jack and plug Three-pin connector that usually carries a balanced audio signal.

Index

Index

Index

M

O

Index

W

warranty (equipment): 24–25, 27–28
WAV files: 57
wow and flutter: 69, 77

X

XLR plug and jack: 48–49, 83, 165–167, 171
XM Satellite Radio: 7, 139, 145–146

Y

Yamaha: 56

Other Books by Robert Harley

The Complete Guide to High-End Audio

Now in a third edition with 640 pages and more than 200 photographs and illustrations, *The Complete Guide to High-End Audio* is the ultimate reference book on high-performance audio! Packed with insider secrets for buying, setting up, and enjoying high-performance audio equipment. Discover for yourself why this universally acclaimed book has sold more than 100,000 copies worldwide in four languages.

"Before you make a mistake, buy Bob Harley's book." —Sam Tellig, *Stereophile*

Call toll-free for your copy
(800) 888-4741
hifibooks.com

Other Books by Robert Harley

Home Theater for Everyone

Robert Harley tells you everything you need to know to become a more knowledgeable buyer of today's advanced home-entertainment systems. In nine fact-filled chapters—covering everything from the basics to technical matters (in plain English!)—this book helps you reap the benefits of Robert Harley's years of reviewing experience, saving you time, trouble, trial and error. From surround-sound to flat-panel televisions, this book has it all!

"[*Home Theater for Everyone*] is the best guidebook we've seen on the subject." —Wayne Thompson, *The Oregonian*

Call toll-free for your copy
(800) 888-4741
hifibooks.com

Critical Acclaim for
Robert Harley's Books

"Brilliant! This book takes Bob Harley's years of reviewing experience and makes it your own. It will accelerate your learning curve and help you get better sound for less money." —John Atkinson, *Stereophile*

"*The Complete Guide to High-End Audio* is a must-have addition to every music lover's library. For less than the price of three CDs, this book is without question the best value in high-end audio." —Shannon Dickson, *The Audiophile Voice*

"I highly recommend this book." —Harry Somerfield, *San Francisco Chronicle*

"This comprehensive, up-to-date coverage of audio fundamentals replaces several reference works as the single best text on high-end audio." —Ed Dell, *Audio Amateur*

"I like this book. I give Harley a lot of credit for tackling a potentially daunting subject with such intelligence and passion." —John Stiernberg, *Sound and Video Contractor*

"Its impressive breadth and depth make it a valuable guide to audio's gold mine. You'll learn valuable things from Harley's book, not just at first reading, but as you come back to it over and over." —Peter Moncrieff, *International Audio Review*

"A monthly magazine like *Stereophile* often devotes space to introductory articles, buying tips, or features on systems and listening rooms; but it could never put together the material in such a structured and consistent way as Robert Harley has achieved in *The Complete Guide to High-End Audio*. . . I can firmly recommend this unique, largely up-to-date book." —Martin Colloms, *Stereophile*

"This is one valuable book, rich in information and insight into the world of high-end audio. I have benefitted immensely from this book and I believe it to be the 'bible' for high-end audio." —Dennis Krishnan, *High-End Magazine*

"Before you make a mistake, buy Bob Harley's book." —Sam Tellig, *Stereophile*

"[*Home Theater for Everyone*] is the best guidebook we've seen on the subject." —Wayne Thompson, *The Oregonian*

"While most books on the subject of home entertainment are out of date before they hit the bookstores, *Home Theater for Everyone* will be useful well into the next century. I highly recommend it. Am I gushing? It's deserved." —Harry Somerfield, *San Francisco Chronicle*